# FIBER COMPOSITES IN MANUFACTURING ENGINEERING AND TECHNOLOGY

# FIBER COMPOSITES IN MANUFACTURING ENGINEERING AND TECHNOLOGY

*By*

*Saroj Shinde*

2016

SBS Publishers & Distributors Pvt. Ltd.
New Delhi

ISBN 13 : 9789380090757

First Published in 2016



Published by:

SBS PUBLISHERS & DISTRIBUTORS PVT. LTD.

2/9, Ground Floor, Ansari Road, Darya Ganj,

New Delhi - 110002,

INDIA

Tel: 0091.11.23289119 / 41563911

Email: mail@sbspublishers.com

www.sbspublishers.com

# Preface

A fiber-reinforced composite (FRC) is a composite building material that consists of three components: (i) the fibers as the discontinuous or dispersed phase, (ii) the matrix as the continuous phase, and (iii) the fine interphase region, also known as the interface. This is a type of advanced composite group, which makes use of rice husk, rice hull, and plastic as ingredients. This technology involves a method of refining, blending, and compounding natural fibers from cellulosic waste streams to form a high-strength fiber composite material in a polymer matrix. The designated waste or base raw materials used in this instance are those of waste thermoplastics and various categories of cellulosic waste including rice husk and saw dust. FRC is high-performance fiber composite achieved and made possible by cross-linking cellulosic fiber molecules with resins in the FRC material matrix through a proprietary molecular re-engineering process, yielding a product of exceptional structural properties. Through this feat of molecular re-engineering selected physical and structural properties of wood are successfully cloned and vested in the FRC product, in addition to other critical attributes to yield performance properties superior to contemporary wood. Advanced composite materials are also known as advanced polymer matrix composites. These are generally characterized or determined by unusually high strength fibers with unusually high stiffness, or modulus of elasticity characteristics, compared to other materials, while bound together by weaker matrices. These are termed advanced composite materials in comparison to the composite materials commonly in use such as reinforced concrete, or even concrete it.

Editor

# Contents

# Chapter 1

# FINITE ELEMENT PROCEDURE FOR STRESS AMPLIFICATION FACTOR RECOVERING IN A REPRESENTATIVE VOLUME OF COMPOSITE MATERIALS

Paulo Cesar Plaisant Junior[1],
Flávio Luiz de Silva Bussamra[*2],
Francisco Kioshi Arakaki[3]

[1]São José dos Campos/SP - Brazil
[2]Instituto Tecnológico de Aeronáutica,
[3]São José dos Campos/SP - Brazil

## ABSTRACT

Finite element models are proposed to the micromechanical analysis of a representative volume of composite materials. A detailed description of the meshes, boundary conditions, and loadings are presented. An illustrative application is given to evaluate stress amplification factors within a representative volume of the unidirectional carbon fiber composite plate. The results are discussed and compared to the numerical findings.

# INTRODUCTION

By analyzing history, it is possible to see the importance of material science applied to Aeronautical Engineering and, in that scenario, the composite materials emerged. As this type of material became more recognized, new branches of research came out. One of these branches is the micromechanics, which is the theory that this study focuses on. Since composite materials play an important role in the modern industry, it is necessary a better understanding of them. Particularly in aeronautical industry, metal alloys have been replaced by composite materials. The best example is the latest Boeing aircraft, 787: 50% of its structure is made of composite materials and 20% of aluminum. Its predecessor has 12% of composite materials and 50% of aluminum (Boeing, 2010).

The analysis of composite materials follows a macro, meso, or micromechanical approach. Micromechanics analyze a laminated plate as a homogeneous anisotropic equivalent plate. In the micromechanical approach, a laminate is modeled as a stacking sequence of homogeneous layers and interlinear interfaces (Ladevèze *et al.*, 2005), therefore the prediction of complex behavior, as delamination, can be assessed (Allix, Ladevèze and Corigliano, 1995). The micromechanical analysis goes down to the constituent properties. The object of study on the micromechanical analysis is the representative volume element (RVE), or unit cell, which is the smallest cell capable of representing the overall response of the unidirectional ply to mechanical and thermal loading (Jin *et al.*, 2008). Figure 1 illustrates an example of RVE.

Jin *et al.* (2008) show the use of three-dimensional finite element models for obtaining stress distribution on composites. Micromechanical finite element models provide data to obtain the micro-stresses at the matrix/ fiber interface, and great benefits can be reached when a micromechanical approach is considered. Micromechanics of failure give a more precise way of composite failure prediction. With the micro-stresses, it is possible to determine the failure initiation in the unidirectional ply (Ha, Huang and Jin, 2008a; Tay *et al.*, 2008; Gotsis, Chamis and Minnetyan, 1998). The material lifetime forecast can be obtained with the use of micromechanics of failure associated with the Accelerated Testing

Method (ATM) and Evolution of Damage (Sihn and Park, 2008; Ha, Huang and Jin, 2008b).

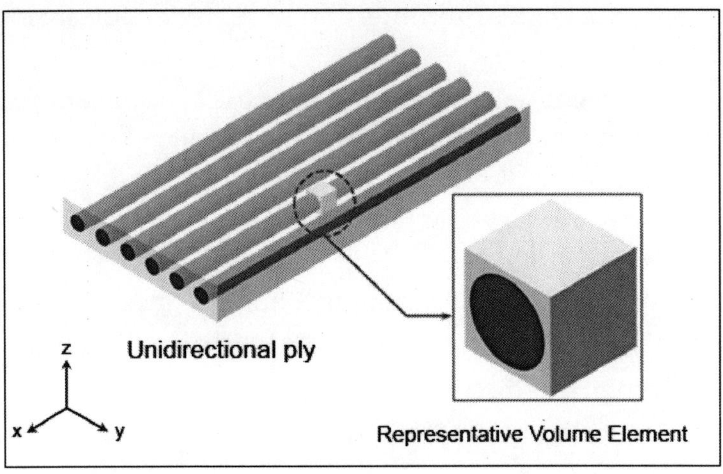


**Figure 1**. Representative volume element in a unidirectional fiber composite play.

The micromechanical theory considers not only the mechanical loads, but also environmental factors, such as thermal loads, as a result of temperature variation and moisture (Hyer and Waas, 2000; Fiedler, Hojo and Ochiai, 2002).

The influence of fiber arrangements on mechanical behavior of laminated pates is discussed by Hojo *et al.* (2009) and Ha, Huang and Jin (2008c). Considerations about the fabrication process are discussed by Aghdam and Khojeh (2003). Studies with reinforcements other than fibers, such as particles, are exposed by Zhu, Cai and Tu (2009). Different types of composites, like bulk metallic glasses (Dragoi *et al.*, 2001), metallic matrix composites (Chaboche, Kruch and Pottier, 1998), and smart composites, which include piezoelectric composites, shape memory alloy (SMA) fiber composites, and piezoresistive composites (Taya, 1999), are also analyzed by micromechanics. Liang, Lee and Suaris (2006) present the results of a comparison between micromechanical finite elements modeling and mechanical testing. RVE use in order to represent the composite material is discussed by Sun e Vaidya (1995), and a study of boundary conditions for the unit cell is shown by Xia *et al.* (2003). From the aeronautical industry to dentistry, a great variety of products can be benefited from a micromechanical study of a composite material. Sakaguchi, Wiltbank and Murchison (2003) show micromechanical studies to predict composite elastic modulus and polymerization shrinkage for dental materials.

The micromechanical theory has direct application for the aeronautical industry. Tsai (2008) points out the use of micromechanics to:

- predict macro mechanical properties (stiffness constants, expansion coefficients);
- control the deformation from mechanical and thermal loads;
- predict a successive ply failure after the first ply failure and;
- adjust empirical data by using micromechanical data.

The objective of this paper is to present and to discuss a methodology in order to obtain the stress amplification factors derived from mechanical and thermal loads in a RVE, with the use of two and three-dimensional finite elements models. Also, this paper aims at accessing stress amplification factors in an orthotropic unidirectional ply with epoxy matrix, carbon fiber, and a perfectly bonded matrix/fiber interface, with 60% of fiber volume fraction. The materials remain in the linear elastic domain. The finite element models are analyzed with the commercial software MSC/NASTRAN, version 70.0.6 (MSC, 2011).

## STRESS AMPLIFICATION FACTORS

In a micromechanical level, there is a difference between the applied and actual stresses within the material, mostly because of the dissimilarity on physical properties of the materials. For example, epoxy matrixes present a lower young modulus than the carbon fiber. When a load is applied to a composite material, due to this stiffness difference, the matrix and the fiber tend to show different stresses, resulting in stress concentrations. Therefore, when a unit load is applied to the material, the stresses within the representative volume are no longer unitary, as shown in Fig. 2.

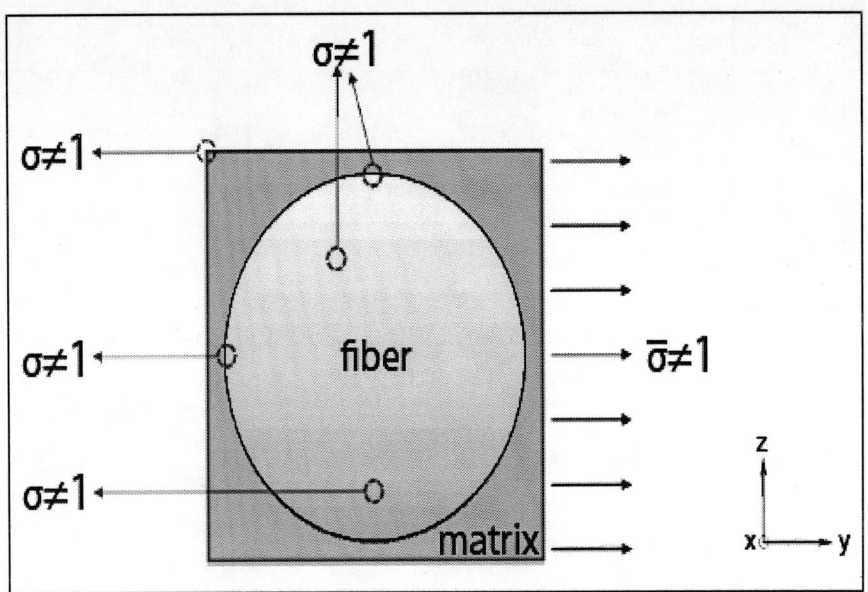


**Figure 2.** Differences between micro and macro stresses.

Jin *et al.* (2008) show that there are amplification factors which relates a uniformly distributed unit load ($\overline{\sigma}$, the macro mechanical load) and the internal micromechanical stresses σ, expressed in Eq. 1:

$$\sigma = \mathbf{M}\overline{\sigma} + \mathbf{A}\Delta T \qquad (1)$$

Where,

M and A are matrices that collect the mechanical and thermal stress amplification factors, respectively, and $\Delta T$ is the increase of the room temperature.

Considering all stress components, referred to the material coordinate system xyz (the same as 123), Eq. 1 can be expanded by Eq. 2 (Jin *et al.*, 2008):

Considering all stress components, referred to the material coordinate system xyz (the same as 123), Eq. 1 can be expanded by Eq. 2 (Jin *et al.*, 2008):

$$\begin{Bmatrix} \sigma_1=\sigma_{xx} \\ \sigma_2=\sigma_{yy} \\ \sigma_3=\sigma_{zz} \\ \sigma_4=\sigma_{yz} \\ \sigma_5=\sigma_{zx} \\ \sigma_6=\sigma_{xy} \end{Bmatrix} = \begin{bmatrix} M_{11} & M_{12} & M_{13} & M_{14} & 0 & 0 \\ M_{21} & M_{22} & M_{23} & M_{24} & 0 & 0 \\ M_{31} & M_{32} & M_{33} & M_{34} & 0 & 0 \\ M_{41} & M_{42} & M_{43} & M_{44} & 0 & 0 \\ 0 & 0 & 0 & 0 & M_{55} & M_{56} \\ 0 & 0 & 0 & 0 & M_{65} & M_{66} \end{bmatrix}_{\sigma} \begin{Bmatrix} \bar{\sigma}_1 \\ \bar{\sigma}_2 \\ \bar{\sigma}_3 \\ \bar{\sigma}_4 \\ \bar{\sigma}_5 \\ \bar{\sigma}_6 \end{Bmatrix} + \begin{Bmatrix} A_1 \\ A_2 \\ A_3 \\ A_4 \\ A_5 \\ A_6 \end{Bmatrix}_{\sigma} \Delta T \tag{2}$$

M can be found by applying unidirectional mechanical loads, one at a time. For instance, if a uniformly distributed unit load is applied at x direction, with no thermal load, Eq. 2 simplifies to Eq. 3:

$$\begin{Bmatrix} \sigma_1 \\ \sigma_2 \\ \sigma_3 \\ \sigma_4 \\ \sigma_5 \\ \sigma_6 \end{Bmatrix} = \begin{bmatrix} M_{11} & M_{12} & M_{13} & M_{14} & 0 & 0 \\ M_{21} & M_{22} & M_{23} & M_{24} & 0 & 0 \\ M_{31} & M_{32} & M_{33} & M_{34} & 0 & 0 \\ M_{41} & M_{42} & M_{43} & M_{44} & 0 & 0 \\ 0 & 0 & 0 & 0 & M_{55} & M_{56} \\ 0 & 0 & 0 & 0 & M_{65} & M_{66} \end{bmatrix}_{\sigma} \begin{Bmatrix} 1 \\ 0 \\ 0 \\ 0 \\ 0 \\ 0 \end{Bmatrix} \tag{3}$$

The resolution of the linear system (Eq. 3) yields Eq. 4:

$$\begin{Bmatrix} \sigma_1 \\ \sigma_2 \\ \sigma_3 \\ \sigma_4 \\ \sigma_5 \\ \sigma_6 \end{Bmatrix} = \begin{Bmatrix} M_{11} \\ M_{21} \\ M_{31} \\ M_{41} \\ 0 \\ 0 \end{Bmatrix} \tag{4}$$

Therefore, the stress amplification factors $M_{11}$, $M_{21}$, $M_{31}$ and $M_{41}$ will be actually the micromechanical stresses $\sigma_1$, $\sigma_2$, $\sigma_3$ and $\sigma_4$, respectively. The same procedure can be applied to all directions. Consequently, the methodology consists of the application of uniformly distributed unit loads to the representative volume model. Thus, the resulting stress at a specific direction gives the corresponding stress amplification factor. Since the stresses at the representative volume vary at each point, the stress amplification factor is not constant.

## MATERIALS

The RVE of a composite material with 60% of fiber volume fraction, subjected to a uniformly distributed load $\overline{\sigma}_2$, is showed in Fig. 3. The fiber is represented as a solid cylinder. The mechanical properties of the matrix and fiber are listed in Table 1, where $E_{ij}$ are Young's moduli, $v_{ij}$ are Poisson's ratios, $G_{ij}$ are shear moduli, and α the thermal expansion coefficient, referred to the material coordinate system xyz (the same as 123).

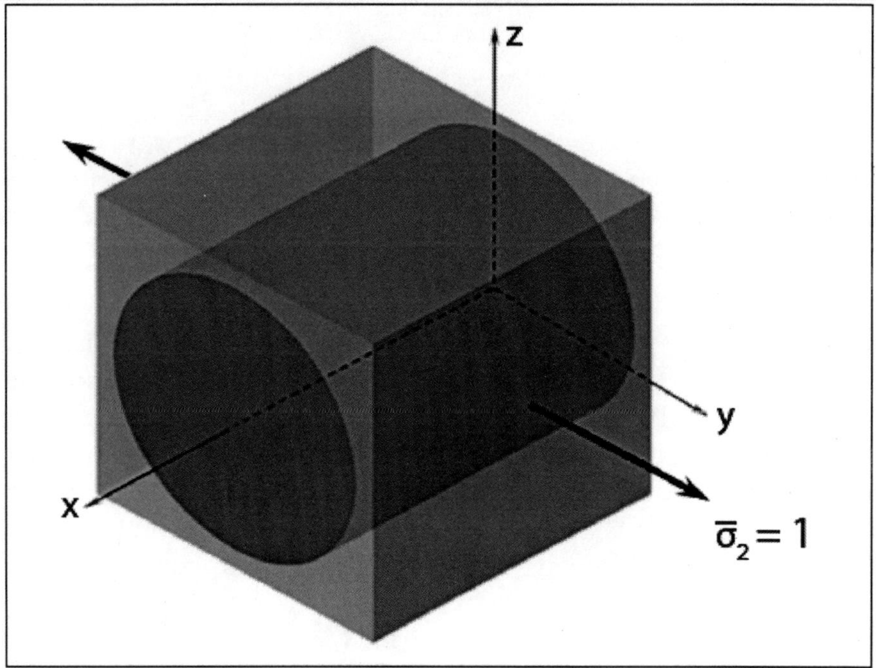


**Figure 3**. Representative volume element of a composite material with 60% of fiber volume fraction.

**Table 1.** Mechanical properties of the representative volume element materials (Think Composites, 2011).

| Carbon fiber | | Epoxy matrix | |
|---|---|---|---|
| $E_{11}$ (Pa) | $2.35\times10^{11}$ | $E_m$ (Pa) | $3.46\times10^{9}$ |
| $E_{22}=E_{33}$ (Pa) | $1.80\times10^{11}$ | $\nu_m$ | 0.35 |
| $G_{12}=G_{13}$ (Pa) | $7.48\times10^{9}$ | $\alpha_m$ ($10^{-6}$/°C) | 57.6 |
| $G_{23}$ (Pa) | $4.90\times10^{9}$ | | |
| $\nu_{12}$ | 0.20 | | |
| $\nu_{13}$ | 0.30 | | |
| $\alpha_1$ ($10^{-6}$/°C) | 0.0 | | |
| $\alpha_2$, $\alpha_3$ ($10^{-6}$/°C) | 8.3 | | |

## NUMERICAL ANALYSIS

Three sets of finite element models are presented and discussed. In the first set, solid hexahedral elements are used to model the matrix and the fiber. The unit cell is represented as a cube with nondimensionalized edge length ($\bar{L} = 1$). Convergence analysis is performed, and stress amplification factors for direct, shear, and thermal loads are presented and discussed. Three-dimensional finite elements are still used in the second set, but the unit cell is no longer modeled as a cube, saving computing efforts. The last set of tests deals with two-dimensional elements. Although they are unable to yield stress concentration factors in transverse directions, in-plane factors are derived. The finite elements solutions are compared with results from Super Mic-Mac software (Think Composites, 2011).

### Three-Dimensional Finite Element Model

The first set of finite element analysis is applied to a three-dimensional model, with the eight-node hexahedral CHEXA Nastran element for the matrix and fiber modeling. The model, including boundary conditions and loads, are further discussed.

## Loads and boundary conditions

When a uniformly distributed tension load is applied to a RVE, the unit cell is constrained at its faces, as shown in Fig. 4, and the free faces must remain flat, as proposed by Jin *et al.* (2008), and Xia, Zhang and Ellyin (2003).

To keep free faces flat, rigid elements are applied to the model. The Nastran rigid element has one master node and one or more slave nodes. The master is the independent one, and it can receive loads. Each slave node will have the same displacements (in the specified direction) of the master one.

Therefore, to keep the right vertical face flat in Fig. 4, a uniformly distributed unit load $\overline{\sigma}_2$ is modeled as a force over only one node. This master node is connected in direction y to the free face nodes (slave nodes) by rigid elements, so all the nodes in this face will have the same displacements in direction y. The nodes in this face must be free at x and z directions (not connected to the master node), in order to move according to the Poisson effect. Rigid body modes are constrained at x=0, y=z=-0.5. Figure 5 shows one model with rigid, spring, and hexahedral elements.

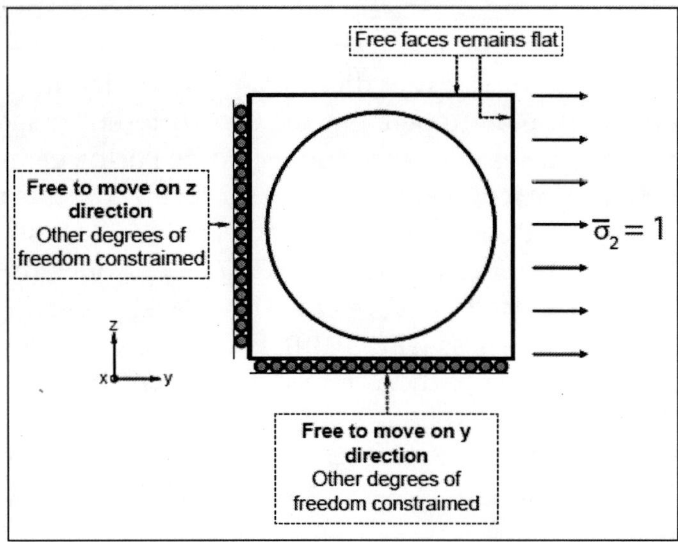


**Figure 4.** Unit cell constraints, subjected to the uniformly distributed unit load $\overline{\sigma}_2$

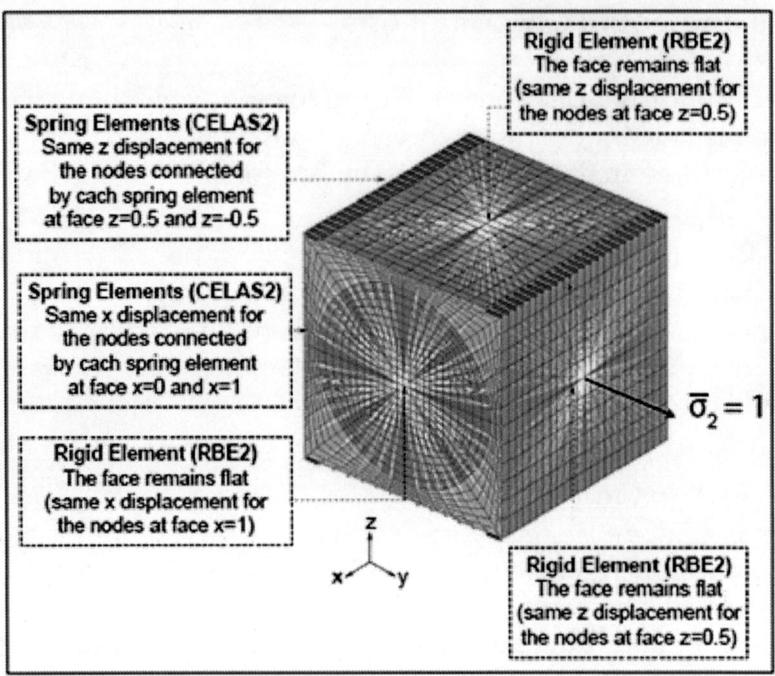


**Figure 5**. Finite element model with: rigid (RBE2), spring (CELAS2) and hexahedral (CHEXA).

The same procedure must be applied to the others free faces. As it is impossible to enforce two dissimilar displacements on a node, nodes cannot figure as dependent on two different master nodes. Therefore, nodes at the edges are connected by spring elements with high stiffness coefficients. For instance, for face z=0.5 it is necessary to connect the rigid element to the nodes on the edges parallel to the x-axis. However, these nodes are dependent ones on the rigid elements from faces y=0.5 and y=-0.5. Thus, Nastran spring element Celas2, with high stiffness (say, 1010), is used to connect only the z displacements for that particular face and, then, this face remains flat. With the Celas2 element, it is possible to connect only one of the degrees of freedom. From faces z=0.5 and z=-0.5, it is connected the z displacements. For faces x=0 and x=1, it is connected the x displacements.

When a uniformly distributed shear load is applied to the RVE, the unit cell is constrained at its faces as shown in Fig. 6, and proposed by Ha *et al.* (2008). The same procedure already explained is applied here. Rigid body modes are constrained.

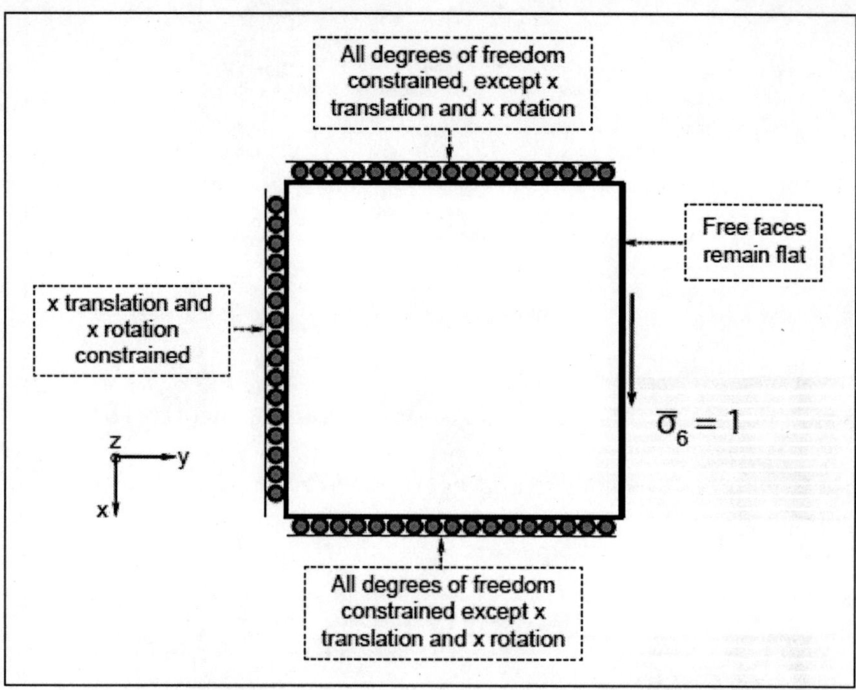


**Figure 6.** Unit cell constraints, subjected to a uniformly distributed shear load $\overline{\sigma}_6 = 1$.

## Convergence analysis

To perform a convergence analysis, a uniformly distributed unit load is applied at direction y. Six finite element meshes are tested (Fig. 7 and Table 2), from a poorly refined 136 element mesh (Mesh 1), to a highly refined 19046 element mesh (Mesh 6).

In order to present the results of the convergence analysis, it is necessary to define a point nomenclature. The points are numbered according to Tay *et al.* (2008), as illustrated in Fig. 8. Points 1, 7 and 4 are the best

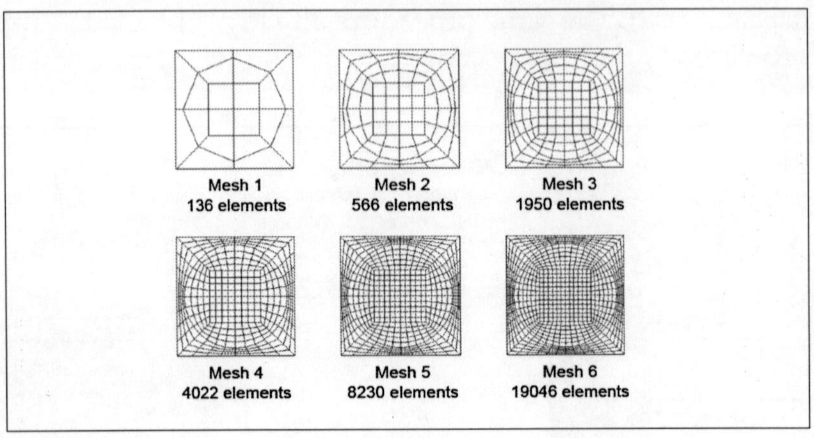


**Figure 7.** Representative volume three-dimensional element meshes.

**Table 2.** Number of elements and degree of freedom (DOF) in three-dimensional models.

| Mesh | Element type | | | Total | DOF |
|---|---|---|---|---|---|
| | CHEXA | CELAS2 | RBE2 | | |
| 1 | 80 | 54 | 2 | 136 | 286 |
| 2 | 480 | 80 | 6 | 566 | 1641 |
| 3 | 1800 | 144 | 6 | 1950 | 5813 |
| 4 | 3840 | 176 | 6 | 4022 | 12171 |
| 5 | 8000 | 224 | 6 | 8230 | 24999 |
| 6 | 18720 | 320 | 6 | 19046 | 57765 |

suitable to have the results displayed, since they are the points of maximum, minimum, and intermediate $\sigma_y$ stresses, respectively. The results of the convergence testing are shown in Figs. 9 and 10.

It can be seen that after 4,000 elements (Mesh 4), the major stress $\sigma_{max}$ at Point 1 starts to converge to the value of 1.5. Above 8,000 elements, the results are in the curve asymptotic portion; therefore the chosen mesh is the number 6. Main stresses at points 1, 4 and 7 are compared with results from the Super Mic-Mac Plus (SMM) software (Think Composites, 2011), which are presented in Table 3. SMM has an extensive database of stress amplification factors resultant from finite element analysis for a wide range of physical properties of fiber and matrix, and different volume fractions. It uses interpolation methods to give the values for specifications, which are not in the database.

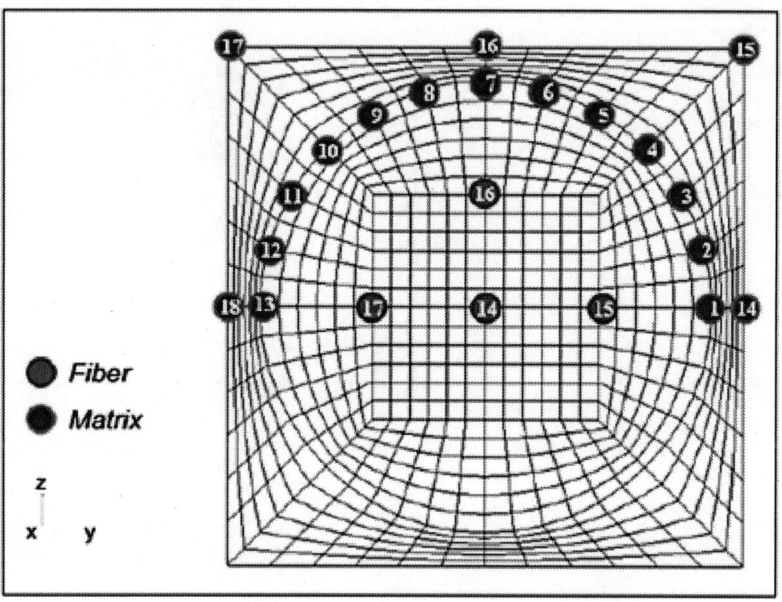


**Figure 8.** Point nomenclature.

The stress distribution $\sigma_y$ (micromechanical) at the RVE is shown in Fig. 11. Since the load is $\bar{\sigma}_2$=1, the stress distribution $\sigma_y$ will characterize the stress amplification factor ($M_{22}$=1.55).

## Stress Amplification Factors for Mechanical Loads

The stress amplification factors with three-dimensional element meshes are found by applying direct and shear uniformly distributed unit loads at the RVE. As already discussed, the stress contours presented in Figs. 12 and 13 represent the distributions of the stress amplification factors. Table 4 shows all the 36 stress amplification factors at Point 1. These results are compared with solutions by SMM in Tables 5 and 6.

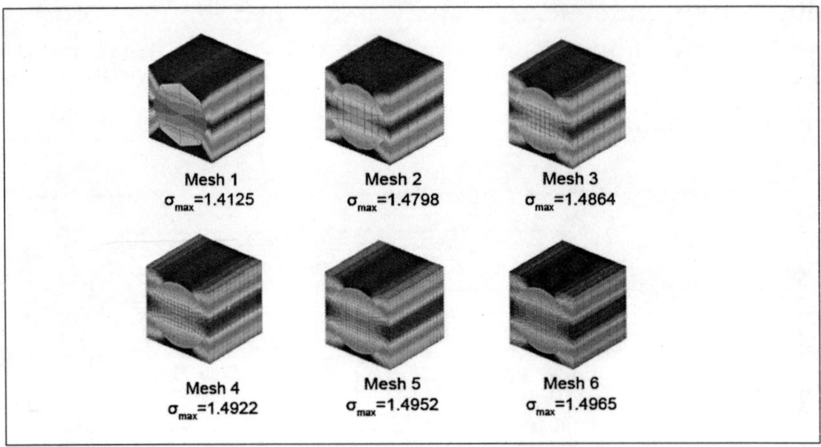


**Figure 9.** Major stress $\sigma_{max}$ at the representative volume element.

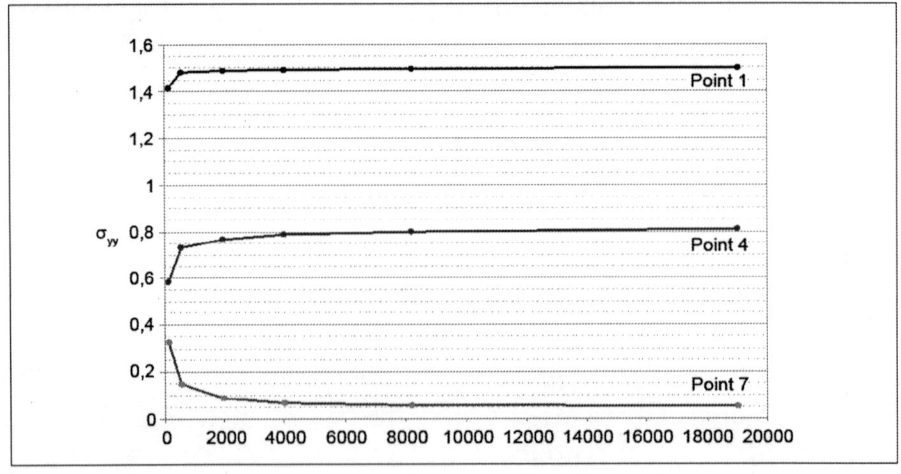


**Figure 10.** Stress convergence in the representative volume element.

**Table 3.** Stress recovering for representative volume elementsubjected to a uniformly distributed load $\overline{\sigma}_2 = 1$.

| Point | 3D model (mesh 6) | SMM | Difference (%) |
|---|---|---|---|
| 1 | 1.49646 | 1.511233 | 0.98 |
| 4 | 0.80874 | 0.810486 | 0.22 |
| 7 | 0.05139 | 0.050587 | -1.59 |

## Stress Amplification Factors for Thermal Loads

Stress amplification factors for thermal loads are obtained by increasing the room temperature by ΔT=1° C. The finite element analysis with Mesh 6, with the same constraints that were previously discussed, yields the stress contours depicted in Fig. 14. Table 7 presents the differences found with SMM.

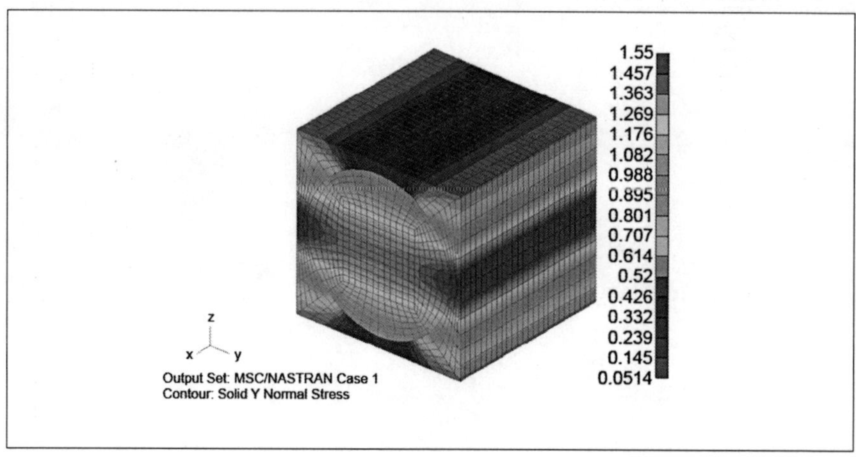


**Figure 11**. Stress amplification factor $M_{22}$.

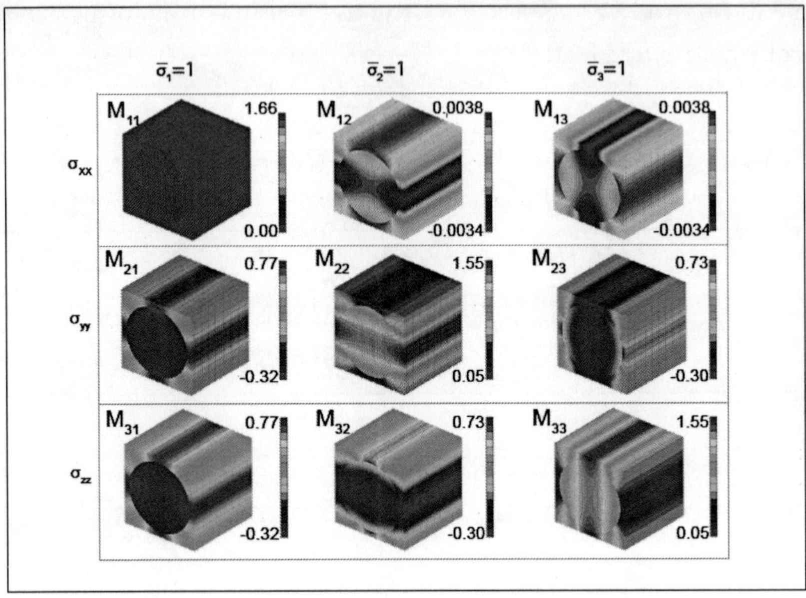


**Figure 12.** Stress contours within the RVE, subjected to macro-mechanical tension load.

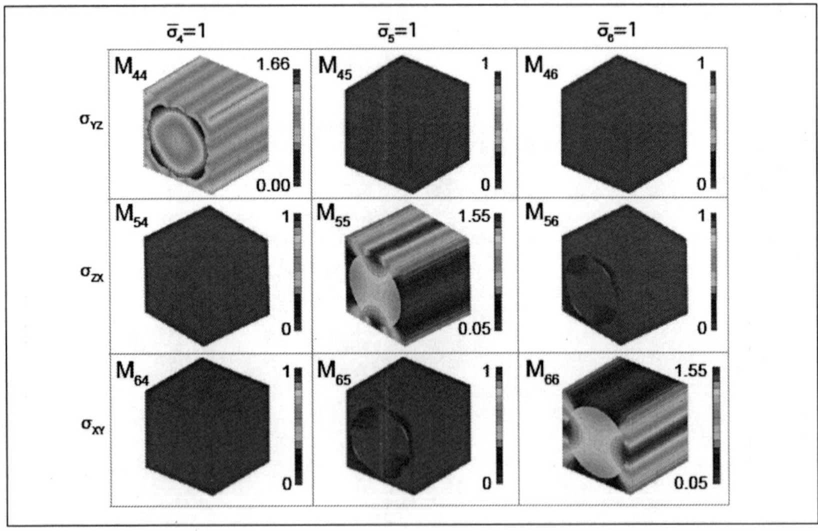


**Figure 13**. Stress contours within the RVE, subjected to macro-mechanical shear load.

**Table 4**. Stress amplification factors from Nastran at Point 1, with Mesh 6.

| Stress | $\overline{\sigma}_1=1$ | $\overline{\sigma}_2=1$ | $\overline{\sigma}_3=1$ | $\overline{\sigma}_{23}=1$ | $\overline{\sigma}_{31}=1$ | $\overline{\sigma}_{12}=1$ |
|---|---|---|---|---|---|---|
| $\sigma_{xx}$ | $2.43\times10^{-2}$ | $7.74\times10^{-1}$ | $-8.00\times10^{-2}$ | $-1.45\times10^{-5}$ | $-2.73\times10^{-9}$ | $-2.20\times10^{-6}$ |
| $\sigma_{yy}$ | $-3.14\times10^{-3}$ | 1.50 | $-2.62\times10^{-1}$ | $1.19\times10^{-5}$ | $-1.25\times10^{-9}$ | $-7.22\times10^{-7}$ |
| $\sigma_{zz}$ | $2.89\times10^{-3}$ | $7.33\times10^{-1}$ | $5.14\times10^{-2}$ | $-5.32\times10^{-5}$ | $-1.32\times10^{-9}$ | $-9.73\times10^{-7}$ |
| $\sigma_{yz}$ | $5.36\times10^{-8}$ | $-7.71\times10^{-6}$ | $3.59\times10^{-6}$ | 1.19 | $-1.07\times10^{-12}$ | $-2.37\times10^{-11}$ |
| $\sigma_{zx}$ | $-2.38\times10^{-10}$ | $1.90\times10^{-10}$ | $-1.37\times10^{-9}$ | $1.26\times10^{-9}$ | $1.60\times10^{-1}$ | $-2.67\times10^{-5}$ |
| $\sigma_{xy}$ | $5.61\times10^{-9}$ | $-4.69\times10^{-6}$ | $8.96\times10^{-7}$ | $-2.94\times10^{-9}$ | $-4.85\times10^{-6}$ | 1.59 |

**Table 5.** Stress amplification factors from SMM at Point 1.

| Stress | $\overline{\sigma}_1=1$ | $\overline{\sigma}_2=1$ | $\overline{\sigma}_3=1$ | $\overline{\sigma}_{23}=1$ | $\overline{\sigma}_{31}=1$ | $\overline{\sigma}_{12}=1$ |
|---|---|---|---|---|---|---|
| $\sigma_{xx}$ | $2.43\times10^{-2}$ | $7.81\times10^{-1}$ | $-8.47\times10^{-2}$ | $-1.15\times10^{-11}$ | $-1.87\times10^{-12}$ | $5.72\times10^{-13}$ |
| $\sigma_{yy}$ | $-3.22\times10^{-3}$ | 1.51 | $-2.75\times10^{-1}$ | $1.95\times10^{-11}$ | $2.34\times10^{-12}$ | $5.10\times10^{-12}$ |
| $\sigma_{zz}$ | $2.89\times10^{-3}$ | $7.34\times10^{-1}$ | $5.06\times10^{-2}$ | $-4.29\times10^{-11}$ | $5.16\times10^{-13}$ | $1.13\times10^{-12}$ |
| $\sigma_{yz}$ | $1.10\times10^{-12}$ | $-5.67\times10^{-11}$ | $1.98\times10^{-11}$ | 1.21 | $9.42\times10^{-14}$ | $1.15\times10^{-13}$ |
| $\sigma_{zx}$ | $-4.05\times10^{-14}$ | $2.24\times10^{-12}$ | $-4.81\times10^{-13}$ | $3.63\times10^{-12}$ | $1.59\times10^{-1}$ | $-2.53\times10^{-11}$ |
| $\sigma_{xy}$ | $2.27\times10^{-15}$ | $-4.80\times10^{-13}$ | $3.34\times10^{-13}$ | $-1.75\times10^{-12}$ | $9.80\times10^{-12}$ | 1.59 |

**Table 6**. Difference between stress amplification factors between Nastran model (Mesh 6) and SMM at Point 1.

| Stress | $\overline{\sigma}_1=1$ | $\overline{\sigma}_2=1$ | $\overline{\sigma}_3=1$ | $\overline{\sigma}_{23}=1$ | $\overline{\sigma}_{31}=1$ | $\overline{\sigma}_{12}=1$ |
|---|---|---|---|---|---|---|
| $\sigma_{xx}$ | 0.00% | 0.86% | 5.65% | - | - | - |
| $\sigma_{yy}$ | 2.62% | 0.98% | 4.82% | - | - | - |
| $\sigma_{zz}$ | -0.14% | 0.16% | -1.59% | - | - | - |
| $\sigma_{yz}$ | - | - | - | 1.69% | - | - |
| $\sigma_{zx}$ | - | - | - | - | -0.57% | - |
| $\sigma_{xy}$ | - | - | - | - | - | -0.15% |

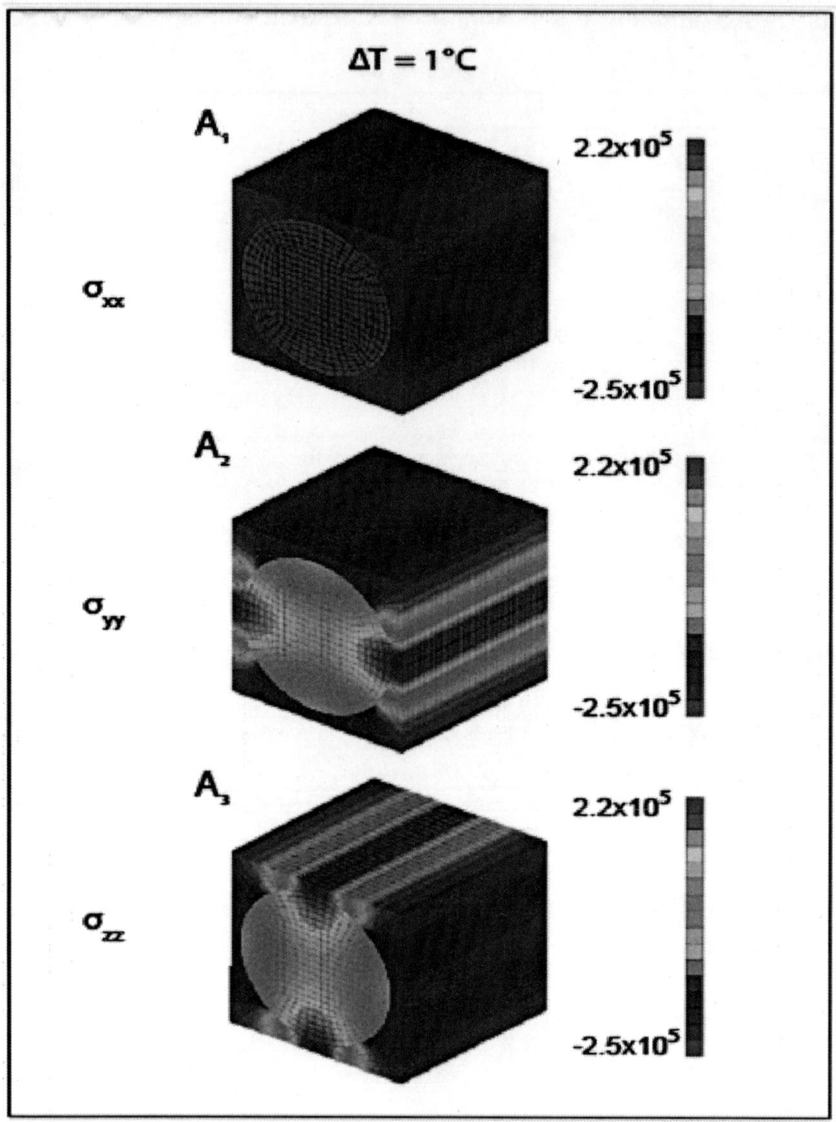


**Figure 14.** Stress contours within the RVE, subjected to thermal load ΔT = 1°C.

**Table 7.** Stress amplification factors at Point 1 for thermal load.

| Stress | Model | SMM | Difference (%) |
|---|---|---|---|
| $\sigma_{xx}$ | $-1.92x10^5$ | $-1.90x10^5$ | -0.74 |
| $\sigma_{yy}$ | $2.06x10^5$ | $2.11x10^5$ | 2.59 |
| $\sigma_{zz}$ | $-1.90x10^5$ | $-1.90x10^5$ | 0.17 |
| $\sigma_{yz}$ | -3.63 | $-1.75x10^{-5}$ | - |
| $\sigma_{zx}$ | $-1.02x10^{-5}$ | $7.89x10^{-7}$ | - |
| $\sigma_{xy}$ | $-5.92x10^{-1}$ | $-1.20x10^{7}$ | - |

## Finite Element Model Simplifications

It is notable that all the results are constant within the element along the x axis, as illustrated in details in Fig. 15. This suggests that: a courser three-dimensional mesh can be used at x direction or the model does not need necessarily to be three-dimensional. A series of model simplifications is carried out. The first part focuses on three-dimensional simplifications by

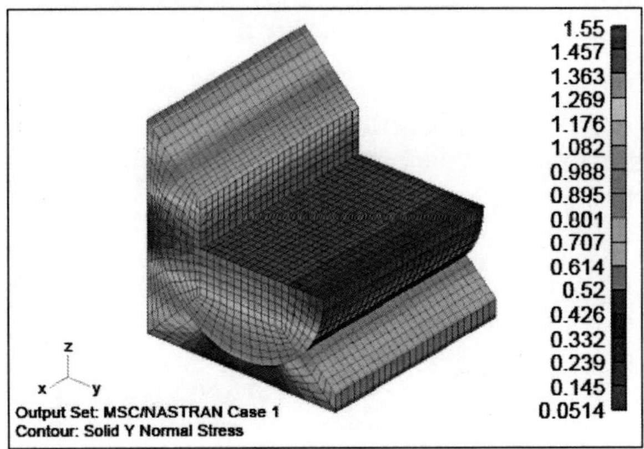


**Figure 15.** Representative volume element in a cutaway view on $\sigma_y$ stress contour. Reducing the number of elements in x direction. The cubic finite element model showed in Fig. 5 is reduced on its half and the resulting model is called Model 1/2. Those model simplifications are carried out up to Model 1/32, as presented in Fig. 16.

The stress amplification factors found with Models 1 to 1/32 are virtually the same, as long as the mesh stays three-dimensional (solid elements). For example, the difference between the full size model and the one with 1/32 of thickness (Model 1/32) is, in the worst scenario, 0.01%.

The next step of this study is to verify the differences between the results of the three-dimensional Model 1/32 and two-dimensional ones. Two different two-dimensional Nastran elements are used: the four-node (linear) quadrilateral membrane CQUAD4 element and the eight-node (parabolic) quadrilateral membrane CQUAD8.

Figure 17 shows $\sigma_y$ stress contours for the RVE subjected to a uniformly distributed unit load $\overline{\sigma_2}$. The correspondent stress amplification factors $M_{22}$ are listed in Table 8. Table 9 shows that two-dimensional models fail in calculating accurate in-plane results.

## CONCLUSION

In this paper, a finite element procedure is presented to access internal stresses in micromechanical analysis of composites materials, with unidirectional fibers. First, three-dimensional finite element models of a RVE are idealized and modeled in Nastran, with CHEXA (hexahedral), CELAS2 (spring), and rigid (RBE2) elements. Convergence analysis is performed. Stress amplification factors are derived within a RVE under tension, shear, and thermal load. Then, two-dimensional Nastran models are also proposed.

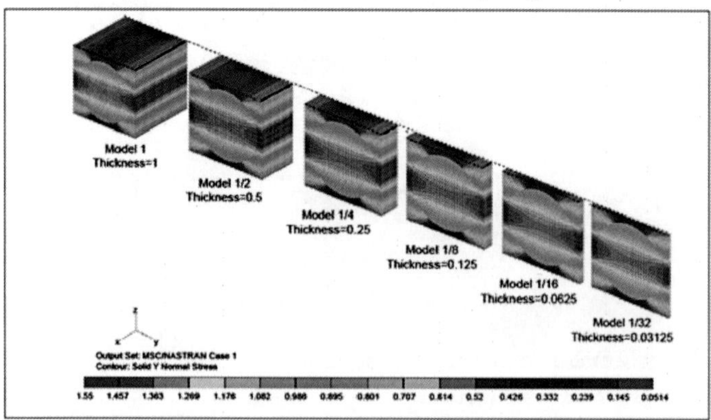


**Figure 16.** Three-dimensional models for RVE, subjected to macro-mechanical unit load $\overline{\sigma_2}$ (σy contours).

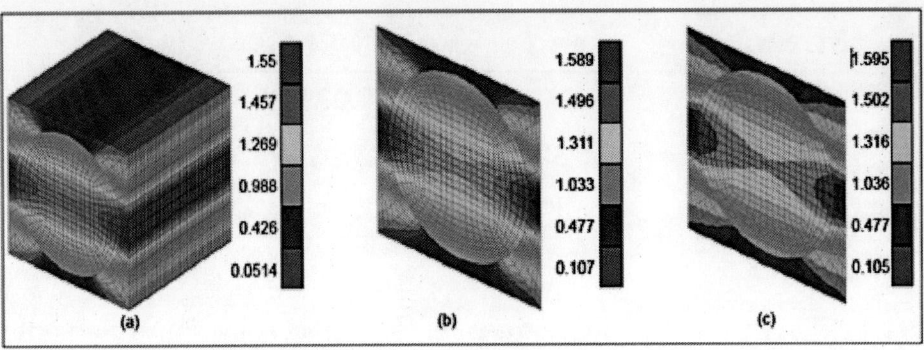


**Figure 17.** Stress amplification factors $M_{22}$ with: (a) three-dimensional Model 1/32 (2984 DOF); (b) linear membrane elements (1464 DOF); (c) parabolic membrane elements (4368 DOF).

**Table 8.** Stress amplification factors with parabolic membrane elements.

| Stress | $\overline{\sigma}_1=1$ | $\overline{\sigma}_2=1$ | $\overline{\sigma}_3=1$ |
|---|---|---|---|
| $\sigma_{yy}$ | 1.545112 | 0.152274 | - |
| $\sigma_{zz}$ | 0.480766 | 0.104816 | - |
| $\sigma_{yz}$ | - | - | 1.287447 |

The presented two-dimensional models fail in providing accurate in-plane stress amplification factors. Good estimation for internal stress amplification factors is achieved with three-dimensional models. Good results

**Table 9.** Differences between stress amplification factors found with Model 1/32 and with parabolic membrane elements.

| Stress | $\overline{\sigma}_1=1$ | $\overline{\sigma}_2=1$ | $\overline{\sigma}_3=1$ |
|---|---|---|---|
| $\sigma_{yy}$ | 3.25% | 41.92% | - |
| $\sigma_{zz}$ | 34.38% | 103.97% | - |
| $\sigma_{yz}$ | - | - | 8.17% |

can be derived with a single transverse layer of solid finite elements, an important feature for nonlinear analyses. The good performance of the presented three-dimensional models shows that good estimates for stress amplification factors can be derived for other than the presented composite (with another volume fraction, or for bidirectional fibers composite), and also to access strain amplification factors.

## REFERENCES

1. Allix, O., Ladevèze P., Corigliano, A., 1995, "Damage analysis of interlaminar fracture specimens", Composite Structures, Vol. 31, No. 1, p. 61-74. doi:10.1016/0263-8223(95)00002-X
2. Aghdam, M., Khojeh, A., 2003, "More on the effects of thermal residual and hydrostatic stresses on yelding behavior of unidirectional composites", Composite Structures, Vol. 62., No. 3-4, p. 285-90.
3. Boeing. 787 Dreamliner - Program Fact Sheet. Retrieved in 2010 August 2, from: http://www.boeing.com/commercial/787family/programfacts.html.
4. Chaboche, J.L., Kruch, S., Pottier, T., 1998, "Micromechanics versus Macromechanics: a combined approach for metal matrix composites constitutive modeling", European Journal of Mechanics, Vol. 17, No. 6., p. 885-908.
5. Dragoi, D., *et al.*, 2011, "Investigation of thermal residual stresses in tungsten-fiber/bulk metallic glass matrix composites", Scripta Materialia, Vol. 45, No. 2, p. 245-52.
6. Fiedler, B., Hojo, M., Ochiai, S., 2002, "The influence of thermal residual stresses on the transverse strength of CRFP using FEM. Composites Part A: Applied Science and Manufacturing", Vol. 33., No. 10, p. 1323-6.
7. Gotsis, P.K., Chamis, C.C., Minnetyan, L., 1998, "Prediction of

Composite laminate fracture: micromechanics and progressive fracture", Composites Science and Technology, Vol. 58, No. 7, p. 1137-49.

8. Ha, S.K., Huang, Y., Jin, K.K., 2008a, "Effects of Fiber Arrangement on Mechanical Behavior of Unidirectional Composites", Journal of Composites Materials, Vol. 42, No. 18, p. 1851-71.
9. Ha, S.K., Huang, Y., Jin, K.K., 2008b, "Life Prediction of Composites using MMF and ATM". In: Tsai, S., 2008, "Strength and Life of Composites", Stanford, Department of Aeronautics & Astronautics of Stanford University.
10. Ha, S.K., Huang, Y., Jin, K.K., 2008c, "Micro-Mechanics of Failure (MMF) for Continuous Fiber Reinforced
11. Composites", Journal of Composites Materials, Vol. 42, No. 18, p. 1873-95.
12. Hojo, M., *et al.*, 2009, "Effect of fiber array irregularities on microscopic interfacial normal stress states of transversely loaded UD-CFRP from viewpoint of failure initiation", Composites Science and Technology, Vol. 69, No. 11-2, p. 1726-34.
13. Hyer, M.W., Waas, A.M., 2000, "Micromechanics of Linear Elastic Continuous Fiber Composites", In: Kelly, A., Zweben, C., 2000, "Comprehensive Composite Materials", Elsevier Science Publishers.
14. Jin, K.K., *et al.*, 2008, "Distribution of micro stresses and interfacial tractions in unidirectional composites", Journal of Composite Materials, Vol. 42, No. 18.
15. Ladevèze, P., *et al.*, 2005, "Micro and meso computational damage modellings for delamination prediction", ICF XI - 11th International Conference on Fracture.
16. Liang, Z., Lee, H.K., Suaris, W., 2006, "Micromechanics-based constitutive modeling for unidirectional laminated composites", International Journal of Solids and Structures, Vol. 43, No. 18-9, p. 5674-89.
17. Msc Software, 2011, "MD Nastran: Integrated, Multidiscipline CAE Solution", Retrieved in 2011 March 23, from http://www.mscsoftware.com/Products/CAE-Tools/MD-Nastran.aspx.
18. Sagushi, R., Wiltbank, B., Murchison, C., 2004, "Prediction of composite elastic modulus and polymerization shrinkage by computational micromechanics", Dental Materials, Vol. 20, No. 4, p. 397-401.
19. Sihn, S., Park, J. W., 2008, "An Integrated Design Tool for Failure and Life Prediction of Composites", Journal of Composites Materials, Vol. 42, No. 18, p. 1967-88.

20. Sun, C. T., Vaidya, R. S., 1996, "Prediction of Composite Properties from a Representative Volume Element", Composites Science and Technology, Vol. 56, No. 2, p. 171-9.
21. Tay, T.E., *et al.*, 2008, "Progressive Failure Analysis of Composites", Journal of Composites Materials, Vol. 42, No. 18, p. 1921-66.
22. Taya, M., 1999, "Micromechanics modeling of smart composites. Composites Part A: applied science and manufacturing", Vol. 30, No. 4, p. 531-6.
23. Think Composites, 2011, "Super Mic-Mac – a new preliminary design tool for composites", Retrieved in 2011 March 23, from http://www.thinkcomposites.com/index_eng.php.
24. Tsai, S., 2008, "Strength and Life of Composites", 1st ed, Stanford: Department of Aeronautics & Astronautics of Stanford University.
25. Xia, Z., Zhang, Y., Ellyin, F., 2003, "A unified boundary conditions for representative volume elements of composites and applications", International Journal of Solids and Structures, Vol. 40, No. 8, p. 1907-21.
26. Zhu, L.J., Cai, W.Z., Tu, S.T., 2009, "Computational Micromechanics of Particle reinforced Composites: Effect of Complex Three-dimensional Microstructures", Journal of Computers, Vol. 4, No. 12, p. 1237-42.

# Chapter 2

# DESIGN WITH ULTRA STRONG POLYETHYLENE FIBERS

Roelof Marissen[1,2]

[1]DSM Dyneema, Urmond, The Netherlands; [2]Delft University of Technology, Faculty of Aerospace Engineering, Delft, The Netherlands.

## ABSTRACT

Ultra strong polyethylene fibers can be made by gel-spinning of Ultra High Molecular Weight Polyethylene (UHMWPE). Such fibers exhibit extraordinary properties. They show very high tensile strength and stiffness and low density. On the other hand, the axial and transverse compression strength is low. This is a large difference with other advanced fibers like glass and carbon fibers. Additionally, the fibers are chemically inert and the bonding strength to other materials like resins is weak. Moreover, the coefficient of friction is very low, so the fiber is extremely slippery. Another property is viscoelasticity; the fiber elongates due to creep at higher loads or temperatures. This exceptional combination of properties explains why gel-spun UHMWPE fibers are not always applied in straight forward ways, e.g. like glass and carbon fibers in composites. On the other hand, weaknesses like the limited compression strength are related to very damage tolerant behavior on a micro scale. This opened application areas like providing of cut resistance. This paper describes some established applications and shows the relationship between the properties and the applications. Furthermore, some emerging applications are discussed and it is demonstrated how weaknesses can be turned into advantages.

## INTRODUCTION

Very strong fibers have found various applications in technology. Glass fibers were about the first non-metallic fibers with strength levels exceeding 2 GPa. Structural application of glass fibers in composites is well established. Such composites are lightweight and high-strength materials. Carbon fibers were developed later and are stronger, stiffer and lighter than glass fibers. Carbon fiber reinforced plastics are superior construction materials with unsurpassed specific strength. Polymer fibers could initially not reach strength levels that are comparable to glass and carbon fibers. However, solvent based spinning technologies enabled the development of ultra-strong polymer fibers.

Two classes of such fibers can be distinguished. One class is based on rigid rod molecules. Well-known products are Kevlar®, Twaron® *or* Zylon®. *The molecular chains of these fibers exhibit some bending stiffness. The other class is made of the very flexible polyethylene molecules. A well-known trade name is* Dyneema® *from DSM. Spectra® is a similar fiber from Honeywell. The interaction between the polymer molecular chains is low for polyethylene. Therefore, very long chains are necessary to provide sufficient load transfer between the macromolecules. Such a polyethylene with long chains is called Ultra High Molecular Weight Polyethylene (UHMWPE). The gel spinning process starts from a high temperature solution of UHMWPE.*

*Cooling causes crystallization from the solution, thus a gel is obtained,* containing polymer crystals and solution. These crystals contain disentangled molecular chains. The disentanglement is conserved during the removal of the solvent. These disentangled chains can be unravelled during a subsequent drawing process. The drawing ratio is very high (about 100) and thus causes extreme orientation of the UHMWPE chains. The parallel oriented chains are again arranged in a crystalline configuration. The crystallinity is high. Demco *et al.* [1]

analyzed various phases with NMR. Roughly summarized about 90% of the fiber material is crystalline. The longitudinal chain orientation "translates" external tension loads to loads on the strong covalent bonds between the carbon atoms of the chains, and this explains the high tensile strength of the fibers. More detailed information on the production and properties of such fibers is described in earlier publications, e.g. by Smith and Lemstra [2], and Jacobs [3]. An early application field that was anticipated during the development of these very strong polymer fibers was the use as reinforcement in composites.

However, this application remains limited due to low axial compression strength of the polymer fibers. Carbon and glass fibers can sustain large axial compression stress if supported by a sufficiently strong and stiff well adhering resin, thus all structural loads can be carried. However, polymer fibers tend to respond with plastic deformation under axial compression. The compression yield strength of rigid rod polymer fibers is typically around 10% of their tensile strength.

The compression yield strength of high strength UHMWPE is even lower and is around 1% of the tensile strength. The explanation is a kind of molecular buckling. This buckling stress is influenced by the low bending stiffness of the molecular chains and by the low interaction between the chains. Consequently, fibers made from rigid rod molecules will show higher compression strength than polyethylene fibers. Rigid rid fibers show more chain-chain interactions, because they contain hydrogen bridges. The Vander Waals bonds and crystalline interactions between UHMWPE chains are much weaker than hydrogen bonds.

Figure 1 shows a Scanning Electron Microscope (SEM) picture of gel-spun UHMWPE fibers with kink bands due to compression loading. The kink bands are the microscopic manifestation of the molecular buckling process. The limited compression strength explains why such fibers are hardly chosen as a reinforcement for structural composites. On the other hand, several investigations e.g. by Marissen *et al.* [4,5] indicate that hybridizing glass or carbon fiber composites with gel-spun UHMWPE fibers can improve the impact resistance, with a small penalty on flexural strength only.

One of the rare applications of composites with gel-spun UHMWPE fibers, without a significant amount of glass or carbon fibers, is for the walls of air cargo panels. The walls are connected to an aluminum frame at the edges of the container and only initial indentation causes real bending stresses and the associated compressive stresses. Larger indentations will cause membrane stresses that are tension by nature. The membrane stresses are transferred to the aluminum corner frame. Consequently, compression strength of the panels is hardly needed. Stiffness and impact strength are required. Air cargo containers are subjected to severe impact loads during their handling on airports. Gel-spun UHMWPE fibers provide the stiffness and excellent impact resistance at low weight. The low weight is desired in view of fuel cost savings and saving of carbon dioxide emissions during flight. Lightness in aviation is of extreme importance. Every kilogram mass saved on flying equipment saves a multitude of kilogram's fuel consumption per year, and accordingly saves carbon dioxide emission. Figure 2 shows examples of such air cargo containers.

**Figure 1.** SEM picture of kink bands in compression loaded gel-spun UHMWPE fibers.

**Figure 2.** Two air cargo containers with panels made from gel-spun UHMWPE fibers and a Turane resin (Courtesy DoKaSch Aircargo equipment GmbH Staudt, Germany).

The colored panels in these containers replace aluminum sheets at about half the panel weight, yet offering about triple impact resistance. Thus a considerable reduction of repair costs is obtained.

Kink bands are common for high strength polymer fibers. However, they are reversible in UHMWPE fibers. They disappear under subsequent tensile loading, without causing noticeable damage. The tensile strength hardly decreases if kink bands were present in these fibers. On the other hand, the compression yielding may be related to the fiber's damage tolerance on a microscopic scale. Glass and carbon fibers behave like elastic rods and bending fracture of those rods occurs if the elastic strength limit of the material is exceeded. High strength polymer fibers will rather show compressive yielding than fracture. The highest toughness may be expected for the fibers with the lowest compression yield strength. Indeed, such effects can be observed. Figure 3 shows a SEM picture of a knot in a single Dyneema® *filament. Extreme curvature and transverse deformation is visible, yet signs of tensile fracture are absent.* Figure 4 shows a

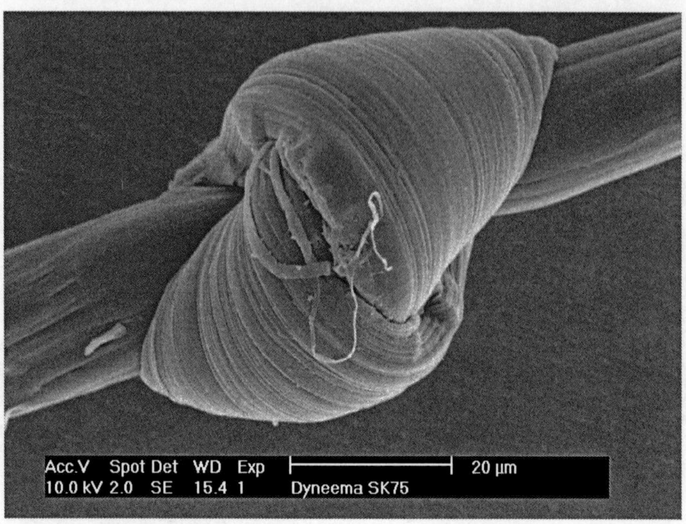


**Figure 3.** SEM picture of a knot in a Dyneema® *filament.*

**Figure 4.** SEM picture of a Dyneema® *filament stretched over the cutting edge of a razor blade.*

Dyneema® *filament that is stretched over the cutting edge of a razor blade.* Again extreme curvature occurs. The filaments deform in the transverse direction and rather spread out over the blade than being cut.

Another special feature of gel-spun UHMWPE fibers is that the melting point is much lower than that of other high strength fibers. Polyethylene is just a low melting polymer. A typical melting point of normal non oriented UHMWPE is about 135°*C. However, the rather perfect* longitudinal orientation of the molecular chains in the crystals causes some increase of the melting point. Typical melting point is about 150°C, it is slightly depending on specific crystal morphologies. Detailed discussions on morphology are presented by Demco *et al.* [1]. Still 150°C is low compared to other high performance fibers. This limits the application area. However, it also offers new unique processing options. Gel-spun UHMWPE fibers can be fused under pressure at a temperature somewhat below the melting point. Such a sintering process enables the production of "matrix free composites" with a slight loss of fiber properties only. Ward *et al.* [6,7] published on the fusion of highly oriented polyethylene fibers. Ward stated that local partial melting of the fibers is important for fusion. His patent mentions a preferred pressure range of about 5 - 20 Bar [7]. However, gelspun UHMWPE fibers can also be fused without noticeable melting. This can be done at much higher pressures. Pressures of more than 100 Bar should than be chosen. The long crystals of the different fibers can be fused under sufficient pressure. Indeed, the low transverse strength of the fibers allows deformation and creation of intense contact between the fibers. A slight rearrangement of molecular chains under pressure may cause fusion of crystals with parallel orientation. Figure 5 presents a SEM micrograph of a cross section showing partly fused fibers being deformed from circular to about hexagon shapes. Some boundaries are still visible; others have disappeared indicating complete fusion.

Polyethylene is "the simplest polymer", it contains only covalent carbon bonds in the chains, and hydrogen atoms as side groups. The absence of other bonds, like ester- or amide bonds implies the absence of properties related to those bonds. So polyethylene (and UHMWPE) fibers are non-polar and insensitive to hydrolysis. This means a good chemical resistance, but also poor bonding to resins and dyes. Some activation of the surface is possible, e.g. with corona or plasma treatments. However, providing very strong bonding to resins, or providing saturated colors to the fibers were as yet unsuccessful.

The above discussion shows that properties and useful technology application of high strength polyethylene fibers is very different from application of other fibers. This paper describes some applications of high strength polyethylene fibers. Firstly, some established applications are briefly presented and discussed in this Chapter. A more extensive presentation of established applications has been presented previously by Vlasblom and Van Dingenen [8].

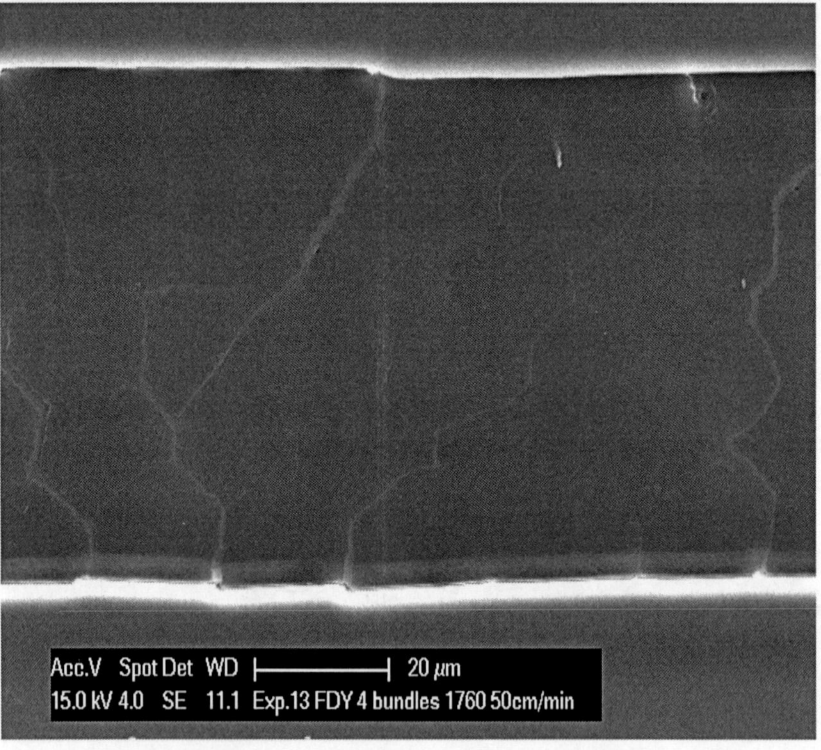


**Figure 5.** SEM picture of fused Dyneema® *filaments.*

# WELL ESTABLISHED APPLICATIONS OF ULTRA

## High Strength Polyethylene Fibers

### *Armor*

Stopping fast projectiles with high strength fibers is well accepted technology. Flexible bullet resistant vests are made from fibrous sheets. Hard armor plates can also be made from fibers. Various slightly different views are possible regarding the physics on the projectile stopping mechanism. The view by Cunniff [9] is chosen here, mainly because it is elementary and straight forward. Cunniff argues that the quality of fibrous armor will be related to the amount of armor that contributes to the projectile catching effect. This contibuting amount increases with increasing sonic speed in the longitudinal direction of the fibers, so with an increasing $(E/\rho)^{1/2}$ value, where $E$ is Young's modulus and $\rho$ is density. Another factor is the amount of deformation energy that can be "consumed" by that contributing amount of material. That energy will be proportional to $(\sigma_{fr}\, \varepsilon_{fr})$, where $\sigma_{fr}$ is the fracture stress and $\varepsilon_{fr}$ is the fracture strain. The final equation by Cunniff is obtained by multiplying the amount of contributing material and the energy consumed by that material:

$$U = \left(\sigma_{fr}\varepsilon_{fr} / 2\rho\right)\left(E/2\rho\right)^{1/2} \qquad (1)$$

where $U$ is an armor performance parameter. Very high strength fibers show approximately linear stress strain behavior, especially for the high loading rates which are typical for ballistics, so $\varepsilon = \sigma/E$ and $\varepsilon_{fr} = \sigma_{fr}/E$. Substitution of this last relationship in Equation (1) yields:

$$U = \sigma_{fr}^{2} \Big/ \left(2\rho^{3/2} E^{1/2}\right) \qquad (2)$$

Equation (2) contains elementary fiber properties only. It shows that a choice of high strength fibers is of highest importance, because strength occurs in the Equation (2) with the highest exponent. Low density is the next important property and low Young's modulus is of some importance for stopping fast projectiles. Table 1 shows some elementary fiber properties. Indeed Dyneema® *fibers exhibit very high strength and low density, thus they are a typical armor material. Of course projectiles* should not be able to travel between the fibers. Therefore fibrous armor contains two perpendicular fiber directions. This reduces the splitting between fibers that would allow easy passing through of projectiles. Hence fibrous armor is made from cross plies of fibers. Figure 6 presents an impression of such armor. Indeed many successful applications of gel-spun UHMWPE fibers armor do exist at present. More details on projectile stopping mechanisms of such armor can be found e.g. in a paper by Jacobs and Van Dingenen [10], and by Van der Werff et al. [11]. Van der Werff also provided perspectives on future performance of fiber-based armor. Laboratory investigations on fiber production by Van der Werff [12] and by Wang et al. [13] with very low polymer concentrations in the initial solution indicate that strength and modulus values could exceed the comercial values with almost a factor two. Equation (2) indicates that the armor performance can be approximately doubled in terms of energy absorption. Note that Van der Werff mentioned a lower number. This is because he argued in terms of projectile speed. The projectile kinetic energy is proportional to the second order of speed. Making such ultimately strong fibers on an industrial scale is quite a challenge. Yet, the argumentation elucidates a future potential.

**Table 1.** Basic fiber properties of a Dyneema® *SK75 yarn with 176* tex.

| Tensile strength [GPa] | Modulus [GPa] | Density [kg/m$^3$] |
|---|---|---|
| 3.4 | 110 | 975 |

Gel-spun UHMWPE fibers are viscoelastic materials. At room temperature, elasticity is dominant under short term loadings (up to days or weeks). However, timedependent behavior is dominant under long-term loading (years). Creep may be considerable under long-term loading, depending on stress and temperature. The molecular origin of the creep deformation is mainly the slip of the individual molecular chains through the crystals. Some chain scission may not be excluded. However, the large creep strains that can be observed (> 50%) can only be explained by chain slip. Indeed, the low interactions (atomic bonds) between the chains and the rather perfect longitudinal crystals with few entanglements can allow such chain slip. Jacobs [3] elaborates extensively on the various aspects of chain slip and the corresponding creep behavior.

The creep of gel-spun UHMWPE fibers is a disadvantage for long-term loaded applications. However, it can advantageously be used for processing. The creep rate increases at high temperatures and it is sufficiently high at temperatures above about 130°C for processing in sufficiently short time. Figure 7 shows two helmet shells made by creep-forming, from a type of cross-ply material as shown in Figure 6. Typically deep drawn helmets of such material show wrinkles due to the 3-D deformation, or they have to be made from a kind of "flower-cut" stack of plies. This cutting provides fiber ends in the helmet and thus may reduce protection. The creep elongation of the fibers allows 3-D draping without wrinkles. The helmets in Figure 7 were made at about 130°C and they are free of wrinkles and all fibers are intact. More details on creep forming can be found in [14].

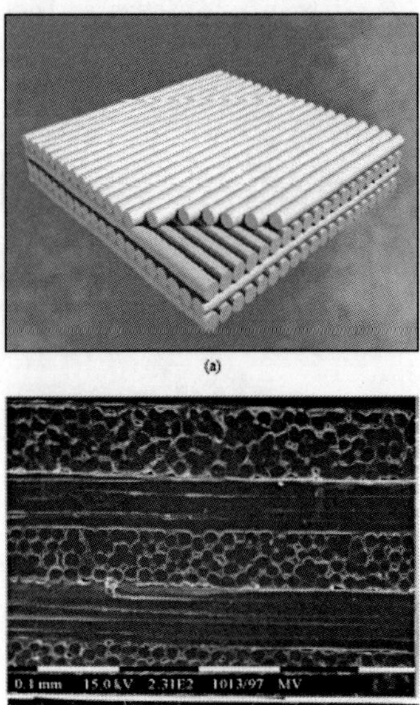


**Figure 6.** Illustration of fiber cross plies for armor, (a) Schematic picture of build up, (b) SEM micrograph of a cross section.

**Figure 7.** Two helmet shells made by creep-forming of gelspun UHMWPE fibers [14].

Another way of making helmets also utilizes the typical features of gel-spun UHMWPE fibers. The possibility to fuse the fibers under pressure at high temperature allows omission of a matrix in "composites" products. This allows dry filament winding without leading the fibers through a resin impregnation bath. It simplifies the set-up considerably and allows fast winding with many yarns simultaneously. Figure 8 shows dry winding around an ellipsoid mandrel with about 30 yarns simultaneously. After completion of the winding process, a hot knife is used to separate the wound ellipsoid shape in two preforms. The fibers melt at the cutting location and thus are welded together, yielding two shaped preforms. Figure 9 shows the "welded" cut line. The resulting halves are stable preforms that can be shaped and consolidated to helmet shells in a hot press. The disadvantage of the low melting temperature is turned here into an advantage, as it allows the simultaneous cutting and fusion of the edges. Thus a practical production of stable preforms is possible.

## Cut Resistance

Gel-spun UHMWPE fibers are difficult to cut. Figure 4 gives an impression on a part of the physics of cut resistance. However, cut resistance in actual products is much more complex. The angle between the blade and the fibers is important as well. Protective textiles are often hybrids with other fibers. Gel-spun UHMWPE fibers are rather stiff as compared to other polymer fibers. Blending with very elastic fibers helps to make products like gloves even more comfortable. Blending may also improve the cut resistance. If the very stiff gel-spun UHMWPE fibers are blended under tension with a stretched elastic fiber, the gel-spun UHMWPE fibers will form loops after relaxation of the tension. Loops remain almost free of stress until the cutting blade has stretched the loop completely.

A striking property of cut resistant gloves made from gel-spun UHMWPE fibers is their comfortable cool feeling. Indeed such fibers have a very high thermal conductivity along the fiber axis. Thus, excessive heat from the hands is easily transferred to the usually colder environment. The combination of cut resistance and cooling is especially attractive in suits for short track skating. The suits offer protection against the sharp skates in case of falling and collision accidents. The suits also provide good transfer of excessive heat that is typically generated in high level exercise. An article by S. Shen *et al.* [15] highlights this behavior on very small scale. Of course, the conductivity is even larger for the perfect crystals that can be made on nano-scale, as compared to industrial scale. However, the less perfect large scale industrial grade of their fibers is already on the market for years, e.g. Dyneema®. *It is not uncommon that ultimate properties are approached on small laboratory scale, where properties associated with large scale industrial production are somewhat lower.*

**Figure 8**. Filament winding with many yarns.

**Figure 9.** Preform with "fiber weld" at the cutting location.

## Ropes, Cables, Wires and Nets

The combination of high tensile strength, stiffness, low density and damage tolerance (as illustrated in the Figures 3 and 4), make gel-spun UHMWPE fibers ideal material for tensile structures that are subjected to frequent handling. Ropes and cables are used in various designs and dimensions, varying from thin lines for kites and fishing lines to very thick cables for offshore mooring. The density of about 0.975 $g/cm^3$ makes cables almost weightless under water, thus they do not have to carry their own weight in deep water mooring.

Parallel fusion of the fibers at increased temperature, but below the melting point allows the production of wire like monolines that are attractive for fishing. Figure 10 shows a SEM picture of such a wire. The individual filaments can still be recognized to some extent. Yet it behaves as a single line. Nets for fishing and fish farming can be made from gel-spun UHMWPE fibers. Again, lightness and small size at high strength are attractive. The bite resistance makes them especially suitable for fish farming. Nets for air cargo pallets show high strength and lightness and thus allow for a reduction of fuel consumption. Gel-spun UHMWPE fibers show low friction coefficients. Measurement results of friction coefficients ($\mu$) yield somewhat varying values. A coefficient of friction $\mu$ = 0.05 is not uncommon. Karuppiah *et al.* [16] investigated the effect of crystallinity on the friction coefficient of UHMWPE. They found reduced friction for increasing crystallinity. Consistently, the very high crystallinity of gel-spun UHMWPE fibers explains the low coefficient of friction. It can be speculated that this is related to the lower mobility of the molecular chains in the crystal. Lower mobility implies lower interaction between the sliding parts. The low friction causes some difficulties regarding connections of ropes and connections to ropes. On the other hand, the relative sliding of filaments allows bending of ropes with little damage. Thus bending fatigue resistance of such cables is large. Smeets *et al.* present some quantitative results on bending fatigue of gel-spun UHMWPE cables [17].

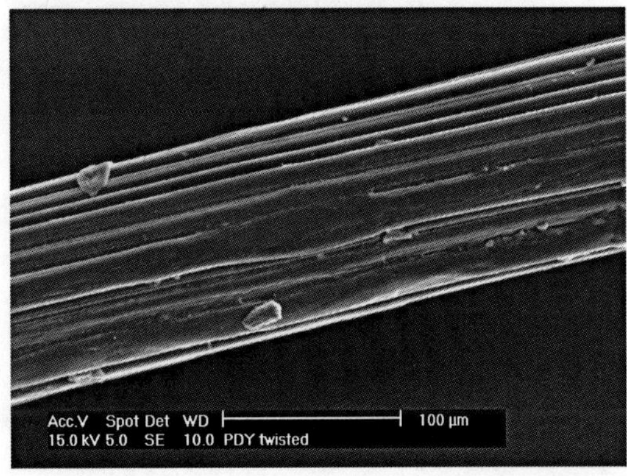


**Figure 10.** Monofilament made by fusing gel-spun UHMWPE fibers.

## RECENT AND EMERGING TECHNOLOGY WITH ULTRA HIGH STRENGTH POLYETHYLENE FIBERS

### Medical Applications

Dyneema Purity® is a very clean grade of gel-spun UHMWPE fiber. It is a fairly recent development, based on a proprietary spinning process. Dyneema Purity® is an ideal material for interactions with the human body, both during demanding surgeries, and over the longer lifespan required for implantable devices. The high strength, softness and abrasion resistance are very valuable assets in demanding applications like sports medicine. The material's low elongation and fatigue resistance offer a superior alternative to traditional materials for surgeons and device manufacturers. Some typical uses include: highstrength sutures, ligament repair, arthroscopic procedures, motion-preserving spinal applications, trauma and surgery of the spine. A striking example is rotator cuff repair. Thousands of rotator cuff repair shoulder surgeries with Dyneema Purity® have been performed already with good results and this surgery is becoming standard practice.

The wear resistance and the possibility for thermal fusion suggest the use of this pure variant as a liner in artificial implantable joints. Such joints may not only provide good wear-resistance, but the fused fibers will also provide high strength and resistance to cracking that is still observed sometimes after long term in vivo use of conventional joint materials. A young company named Cartificial explored this field. The results were very encouraging. Unfortunately, testing of medical devices is very expensive and this start-up company did run out of cash before the commercial potential was established sufficiently and the company was discontinued. Yet, the potential might be materialized later.

## Kite Based Wind Energy

Wind turbines provide an increasing part of electric energy production. A much lighter alternative for wind turbines are kites. Although many practical problems have to be solved, kites are potentially more powerful and more effective energy generators than conventional wind turbines. Ockels is one of the pioneers of this concept; see e.g. Podgaets and Ockels [18]. His kite based ladder mill design should generate approximately 100 MW. Many concepts for kite based energy generation have already been proposed. All kite based energy generation systems will require lightweight cables, allowing much "handling". The high tenacity (specific strength) and damage tolerance of gel-spun UHMWPE fiber will make it a first choice material for the required cable systems.

## Connections

The low friction coefficients of gel-spun UHMWPE fiber were discussed before. The low coefficient of friction makes gel-spun UHMWPE fibers notorious for knot slippage. The friction coefficient of fibers against steel is about 0.1. The fiber-fiber friction coefficient is about 0.05 only. The behavior of knots suggests that these numbers may even decrease under high normal forces. Strong knots in gel-spun UHMWPE fibers require additional loops. A square knot that provides good hold in conventional fibers behaves as a "large force sliding knot" in gel-spun UHMWPE fibers. This may be impractical for standard connections. However, it may be turned into a great advantage. A square knot can be used to connect parts with a loop. Subsequently, the loop can be tensioned by a large mechanical force at the end of the fibers. The strong fibers will allow that large force. About 10% of this force will be transferred through the square knot to the loop, thus tensioning the connected parts considerably. Most of this tension will remain after removing the force at the ends. An additional square knot (or optionally more knots) will cause a durable fixation. Variations of this procedure are under development for surgery, allowing for a safer procedure involving temporary fixation, optionally adjustment, and final fixation.

A variation of this slip-block technology is the use of auxiliary features. A system that is easily tensioned and provides excellent holding power is based on the use of small metal rings. An assembly of two or three rings can be used.

Figure 11 shows a schematic picture of a clamping method with a three ring clamping system. The dimensions are optimized for a Dyneema® *cable of 1400* tex (1400 gram/kilometer). The upper ends in the figure are the tensioning ends. The lower ends are connecting ends, e.g. parts of a loop to be tensioned. The high strength of the fibers, together with the low friction, allows firm tensioning. On the other hand, the loading force at the connecting end presses the rings together and thus prevents reversed sliding. The extra loop configuration has the best holding force, about 3000 N for the above mentioned cable; the standard winding pattern allows easier and faster tensioning. Many variations are possible. All have in common that the tensioning ends are between the rings and the loaded ends are around the rings. Figure 12 shows photographs of the assembled rings with the fibers. Figure 13 shows a variation with two rings.

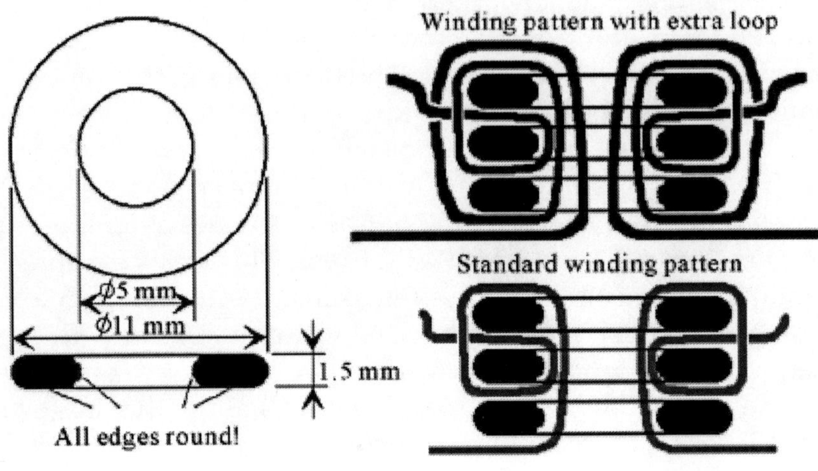


**Figure 11.** Schematic presentation of ring clamping.

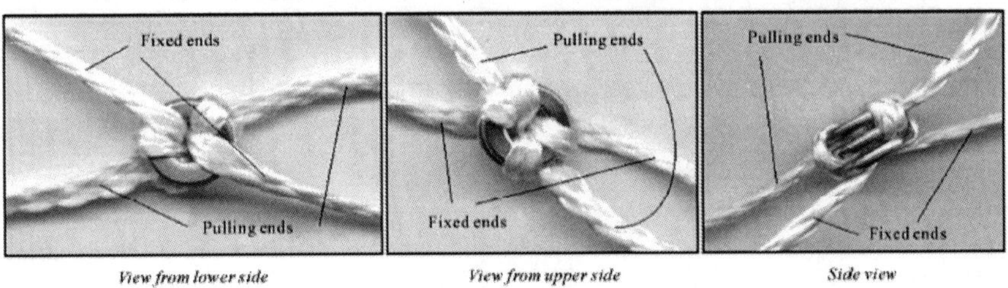


**Figure 12.** Pictures of the installed three-fold ring clamp with extra loop.

Connections may be desired to be permanent, but sometimes it is important that quick release and connection is possible. Metal shackles, or carabiners are used in such cases. However, they are heavy and hard. A metal shackle at the end of a swinging rope can be a hazard. A recent alternative is a soft shackle. Soft shackles are typically made of gel-spun UHMWPE fibers. Reasons for this choice are the high tensile strength and damage tolerance. Another reason is that the slippery character of such fibers allows easy use. Opening can even be done easily after high loading. The special shape still provides a good locking behavior if closed and loaded. Figure 14 shows a LIROS XTR soft shackle in opened and closed condition. Colligo Marine offers a similar "softie" with an additional feature for keeping it closed. Such soft shackles are light, strong and practical in use.

## Hinges

Connecting rigid structures with cables may allow for some flexibility. The connection can be designed in such a way that the flexibility is optimized. This will create a kind of cable hinge, or in general a fiber hinge. The fibers may be present as yarns, cables, or fabrics, depending on the specific design. The use of gel-spun UHMWPE fibers in such hinges is especially advantageous. A rough but effective way of making a line-hinge is to make a composite plate of gel-spun UHMWPE fibers, e.g. by impregnating a fabric with a resin and curing the resin, followed by folding the plate. Indeed, it will not break completely, only the resin breaks! This is unlike other composites and it is attributed to the typical fiber properties. Folding it a few times in both directions and pressing the fold line creates a strong and flexible linehinge in the plate. Another way of making a hinge is shown in Figure 15. If made from Dyneema Purity® and a surgical steel quality, it may potentially be used in arthroplasty as a strong artificial finger joint. Figure 16 shows a possible joint with the kinematics resembling that of a ball bearing. This could be an artificial hip joint that does not generate wear particles. Wear particles are the cause for loosening of the stem-femur connection on long term for conventional implants. The four cables prevent relative translation of both parts, but allow rotation. The body weight will be carried mainly by the lower able. This cable may be designed thicker than the other ones. The extreme tensile strength of Dyneema Purity® and the possibility to apply rather thick cables allow overdesign of the critical cable up to load carrying capacity of a few tons, whereby creep will be virtually eliminated.

## Fiber Modifications

Some "weaknesses" of UHMWPE fibers may even be enhanced. Ropes made of UHMWPE fibers allow frequent bending on winches and sheaves. A lubricating coating on the fibers may even enhance this effect. A so called bending optimized fiber was developed this way. Details are presented by Smeets [17]. Cables made from this modified fiber show even further improved bending fatigue properties.

Also the production of gel-spun UHMWPE fibers allows some flexibility. Pigments and other functional constituents can be incorporated into the fibers during fiber production. A new implantable blue fiber "Dyneema Purity® BLUE" was recently developed and presented at the Medical Design and Manufacturing East Conference & Exposition (MD&M East) 2010 [19]. Blue fibers provide improved visibility for surgeons, due to improved contrast to body tissue.

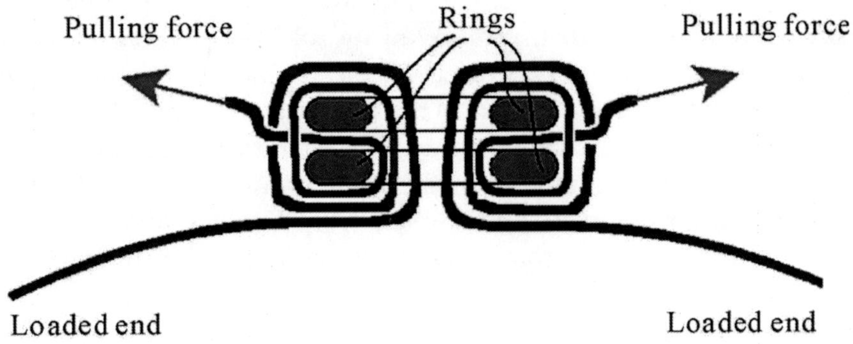


**Figure 13.** Ring clamp variation with two rings.

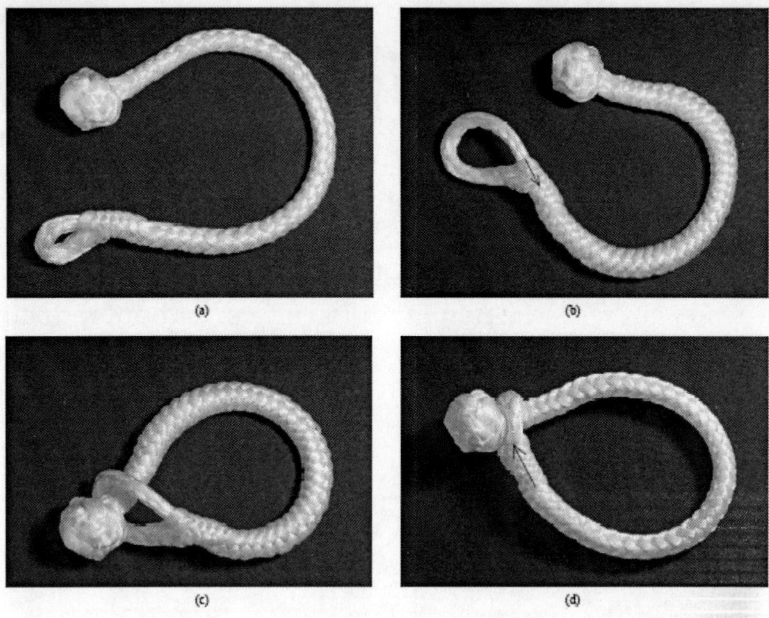


**Figure 14.** Closing a Liros soft shackle; a) Initial condition, b) Opening the loop, c) Feeding the knot through the loop, d) losing the loop.

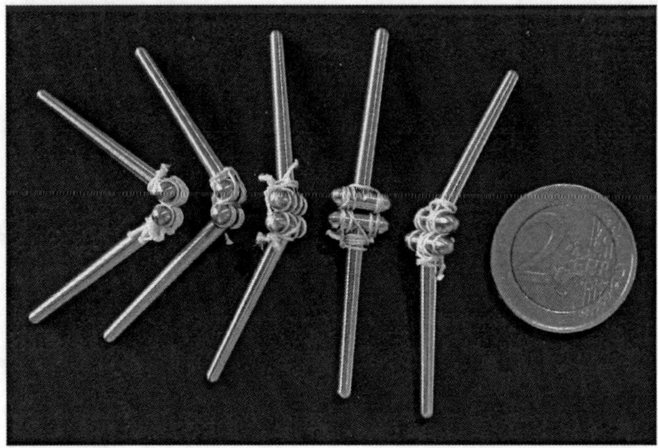

**Figure 15**. Set of fiber-hinges with one degree of freedom; rotation around one axis. The size as compared to a 2 Euro coin illustrates the possibility of design as a potential finger joint implant.

Another modification of gel-spun UHMWPE fibers is the incorporation of short thin mineral fibers in the gel-spun filaments. Figure 17 shows a SEM micrograph of gel-spun UHMWPE fibers containing a mineral fiber. The mineral fiber is made visible (lighter grey scale) by using back scattered electrons. So far, this technique is only published in patents [20]. The commercial launch was at Expoprotection in Paris (France) at 4 November 2010. These fibers with incorporated mineral fibers exhibit about twice the cut resistance of the already cut resistant non modified fibers, without reducing the wearing comfort. The constitution of the mineral fibers is such that they will dissolve in the human body, in case they would be released from the polyethylene filaments and would enter the human body.

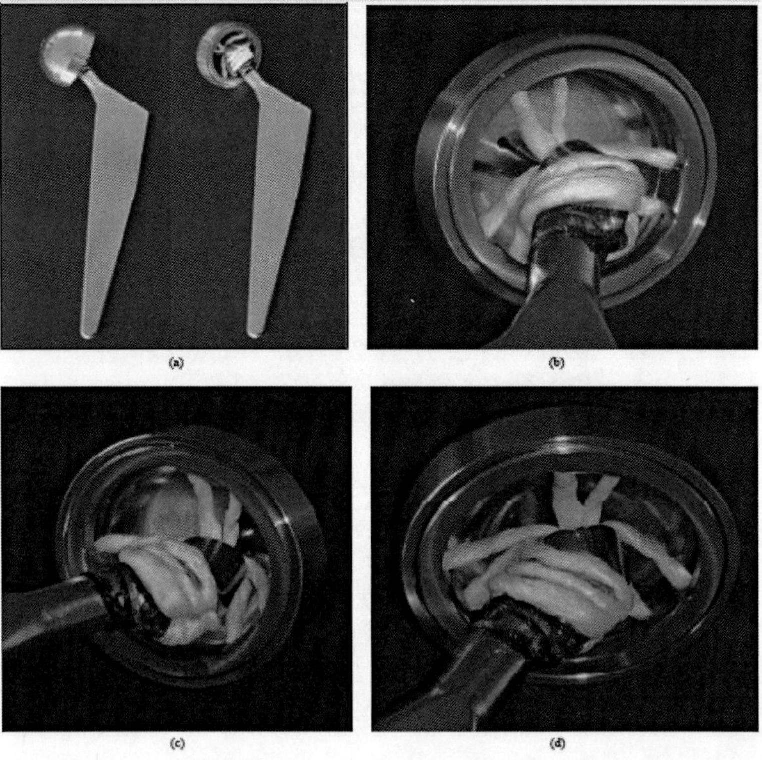


**Figure 16.** Fiber-hinge with rotational freedom in various directions. Designed as a potential hip joint implant, (a) View showing the head rotated in two different positions, (b)-(d) Various views on the inside showing three white cable pairs to the "equator" and one black cable to the "pole" of the cup.

The above innovations demonstrate that new future variants of gel-spun UHMWPE fibers are possible and may become available, allowing even more applications.

## DISCUSSIONS

Gel-spun UHMWPE fibers already have numerous applications in a wide variety of technology fields. However also many application trials failed because of structural limitations of these fibers. This is in contrast to the situation with glass or carbon fibers which are extensively applied in composite materials, but application does not extend much beyond this technology field. This creates an almost paradoxical situation: Fibers with most "complete" properties (tension, compression and transverse strength) are applied in a rather narrow field (composites), whereas gel-spun UHMWPE fibers with "incomplete" properties (tension strength only, weak in compression and in transverse direction) have applications in many different fields. This paradox is discussed in some detail below and resolved to some extent.

**Figure 17.** Gel-spun UHMWPE fiber filaments with incorporated short mineral fibers.

Johan Cruijff once stated (in Dutch): "Every disadvantage comprises an advantage". This wisdom was related to soccer, but it also applies to gel-spun UHMWPE fibers. Such fibers could roughly be qualified as "having all properties but one being negligible". The not-negligible property is that it allows application of extreme tension loading. This character is due to the uniaxial molecular structure of such fibers.

However, the lack of other properties can indeed be an advantage for some applications. The extreme "micro toughness" as illustrated in the Figures 3 and 4 is related to the limited transverse stress transfer in the filaments, thus the tensile stresses are equalized over the filament cross section. High local peak stresses in the fibers are thus annihilated. Of course, it is important that the response in the weak directions is deformation and not material separation, so in fact not all non-tension properties are low. For example, the fracture strain in compressive and transverse direction is high. So weaknesses in those directions are only apparent in terms of force.

It was demonstrated above that the high tenacity, together with disadvantages like creep deformation, a low friction coefficient, and low melting point can all be turned into an advantage, for specific applications, enabling the design of products or processes that show properties with some unique aspects. Finding such possibilities may require creativity but when found, they often lead to unique advantages. Some opportunities for such special products or processes are presently known. However, still a challenge remains. There is no reason to assume that possibilities for new attractive designs with such exceptional fibers are exhausted. Moreover, the gel-spinning process is versatile, and fibers optimized for specific applications, like ballistic protection, surgery, or cut resistance can be (further) developed.

## REFERENCES

1. D. E. Demco, C. Melian, J. Simmelink, V. M. Litvinov and M. Möller, "Structure and Dynamics of Drawn Gel-Spun Ultrahigh-Molecular-Weight Polyethylene Fibers by 1H, 13C and 129XE NMR," *Macromolecular Chemistry and Physics*, Vol. 211, No. 24, 2010, pp. 2611-2623. doi:10.1002/macp.201000455
2. P. Smith and P. J. Lemstra, "Ultrahigh Strength Polyethylene Filaments

by Solution Spinning/Drawing," *Journal of Materials Science,* Vol. 15, No. 2, 1980, pp. 505- 514.doi:10.1007/BF02396802

3. M. J. M. Jacobs, "Creep of Gel-Spun Polyethylene Fibres," Ph.D. Thesis, Eindhoven University of Technology, Eindhoven, December 1999.
4. R. Marissen, L. Smit and C. Snijder, "Dyneema Fibers in Composites, the Addition of Special Mechanical Functionalities," *Conference Proceedings, Advancing with Composites* 2005, Naples, October 2005.
5. J. G. H. Bouwmeester, R. Marissen and O. K. Bergsma, "Carbon/ Dyneema® Intralaminar Hybrids: New Strategy to Increase Impact Resistance or Decrease Mass of Carbon Fiber Composites," *ICAS2008 Conference Anchorage,* Alaska, September 2008.
6. I. M. Ward and P. J. Hine, "The Science and Technology of Hot Compaction," *Polymer,* Vol. 45, No. 5, 2004, pp. 1413-1427. doi:10.1016/j.polymer.2003.11.050
7. I. M. Ward, P. J. Hine and K. Norris, "Polymeric Materials," Patent US6277773, August 2001.
8. M. P. Vlasblom and J. L. J. van Dingenen, "The Manufacture, Properties and Applications of High Strength, High Modulus Polyethylene Fibers," In: A. R. Bunsell, Ed., *Handbook of Tensile Properties of Textile and Technical Fibres,* Woodhead Publishing Ltd., Cambridge, 2009. doi:10.1533/9781845696801.2.437
9. P. M. Cunniff, "Dimensionless Parameters for Optimization of Textile-Based Body Armor Systems," *Proceedings of* 18*th International Symposium on Ballistics, San Antonio,* November 1999.
10. M. J. N. Jacobs and J. L. J. van Dingenen, "Ballistic Protection Mechanisms in Personal Armour," *Journal of Materials Science,* Vol. 36, No. 13, 2001, pp. 3137-3142. doi:10.1023/A:1017922000090
11. H. van der Werff, U. Heisserer and S. L. Phoenix, "Modelling of Ballistic Impact on Fiber Composites," Personal Armour Systems Symposium 2010, Quebec City, September 2010.
12. H. van der Werff and A. J. Pennings, "Tensile Deformation of High Strength and High Modulus Fibers," *Colloid & Polymer Science,* Vol. 269, No. 8, 1991, pp. 747-763. doi:10.1007/BF00657441
13. J. Wang and K. J. Smith, "The Breaking Strength of Ul tra-High Molecular Weight Polyethylene Fibers," *Polymer,* Vol. 40, No. 26, 1999, pp. 7261-7274. doi:10.1016/S0032-3861(99)00034-8
14. R. Marissen, D. Duurkoop, H. Hoefnagels and O. K. Bergsma, "Creep-Forming of High Strength Polyethylene Fiber Prepregs for the Production of Ballistic Protection

15. Helmets," *Composites Science and Technology*, Vol. 70, No. 7, 2010, pp. 1184-1188. doi:10.1016/j.compscitech.2010.03.003

16. S. Shen, A. Henry, J. Tong, R. T. Zheng and G. Chen, "Polyethylene Nanofibres with Very High Thermal Conductivities," *Nature Nanotechnology*, Vol. 5, No. 4, March 2010, pp. 251-255. doi:10.1038/nnano.2010.27

17. K. S. K. Karuppiah, A. L. Bruck, S. Sundararajan, J. Wang, Z. Q. Lin, Z. H. Xu and X. D. Li, "Friction and Wear Behavior of Ultra-High Molecular Weight Polyethylene as a Function of Polymer Crystallinity," *Acta Biomaterialia*, Vol. 4, No. 5, 2008, 1401-1410. doi:10.1016/j.actbio.2008.02.022

18. P. J. H. M. Smeets, M. P. Vlasblom and J. C. Weis, "Latest Improvements on HMPE Rope Design for Steel Wire Rope Applications," *Proceedings, OIPEEC* 2009, *3rd International Ropedays*, Stuttgart, March 2009.

19. A. R. Podgaets and W. J. Ockels, "Laddermill Sail: A New Concept in Sailing," *International Conference on engineering Technology, ICET* 2007, Kuala Lumpur, 11-14 December 2007.

20. "DSM Dyneema Launches Dyneema Purity® BLUE," http://www.dyneema.com/en_US/public/dyneema/page/n ewsitems/PurityBLUE.htm

21. R. Marissen, E. F. F. de Danschuttter and E. Müller, "Cut Resistant Yarn, a Process for Producing the Yarn and Products Containing the Yarn," Patent WO2008046476, April 2008.

# Chapter 3

# EFFECTS OF RC BEAMS REINFOCEMENT USING NEAR SURFACE MOUNTED REINFORCED FRP COMPOSITES

Slobodan Rankovic[1], Radomir Folic[2],and Marina Mijalkovic[1]

[1]The Faculty of Civil Engineering and Architecture of Niš, Serbia epartment of Civil Engineering of the Faculty of Technical Sciences, Novi Sad, Ser

[2]Department of Civil Engineering of the Faculty of Technical Sciences, Novi Sad, Serbia

## ABSTRACT

This paper analyzes application of modern reinforcement methods for reinforced concrete (RC) beams using fiber-reinforced polymer (FRP) materials. Basic characteristics of FRP materials and the method of mounting the FRP bars within concrete, that is, near the surface of the beams (NSM method) are presented. The properties of this method and its advantages in comparison to externally bonded reinforcement laminate method (EBR) have been analyzed.

The results of measured deflections and width of the cracks of the beams reinforced by FRP bars, depending on the load are presented and discussed, in comparison to the results obtained from the non-reinforced beams. The experimental research was published at the Faculty of Civil Engineering and Architecture of Niš in 2009.

## INTRODUCTION

Application of composite materials has become very prominent lately, especially for remedying and reinforcing the reinforced concrete (RC), pre-stressed concrete (PC), ma- sonry, timber and even various steel structures. Usage of fiber reinforce polymer (FRP) composites as additional reinforcement, is a very attractive technique due to numerous advantages, in respect to the conventional methods (addition of new concrete and steel reinforcement, pre-stressing, addition of new steel elements), particularly for RC structures reinforcement. The most important advantage of these composite materials is their high strength and lightness, resistance to corrosion and simplicity in installation [7]. In combination with pre-stressing they give especially good results, and they are particularly suitable for applications in seismic areas. The progressively increasing usage of FRP is illustrated by the fact that numerous global manufacturers of material, e.g. from Switzer- land (Sika), Italy (Mapei, Sinit, Sireg), USA (Hughes Brothers), Canada (Pultral), Japan..., produce some of the varieties of these products [1]. Even though the initial investment is higher, due to a higher price of the material, the speed and ease of installment, and re- distance to the aggressive environment, retained dimensions of the structures and esthetic appearance, they can be advantageous in respect to the other ways of strengthening [9].

The non-metal reinforced used, most often produced by pultrusion process, whose basic component (polymer fibres) can be made of carbon (CFRP), glass (GFRP) or aramid (AFRP). The other component is the matrix, most frequently made of epoxy resin, with the filler and additives for enhancement of some of the properties of the final product, FRP reinforcement. [6]. Then the reinforcement is shaped into laminates, sheets, bars or strips. (figure 1).

**Figure 1**. Shapes of FRP elements: Laminates, sheets, bars.

They are used in reinforcing the structures of concrete, bricks, metal and timber against buckling, shearing or compression. Various structural elements can be reinforced: beams, slabs, walls or columns (figure 2).

**Figure 2.** Examples of application of NSM FRP bars (Nanni 2000).

When reinforcing the RC structures exposed to buckling, two primary methods of ap- plication of FRP stand out: EBR - external bonded laminate reinforcement and NSM – near surface mounted bars. As opposed to the EBR method of using the FRP elements, which has been developing for more than two decades, the NSM method of FRP bars ap- plication emerged only in the last decade and has not been sufficiently dealt with in lit- erature, and is lacking appropriate recommendations and designing guidelines. Interna- tionally, the fundamentals for design and application of these structural reinforcement systems were laid out in the papers by Nanni, Rizkalla, Teng, De Lorenzis, Park, Jung, Barros, Derby-a, Thiagarajan, Ashour, Wang, Belarbi, Galati and others, and in Serbia by R. Folić and S. Ranković [7,8,9,10,11].

## FRP MATERIAL CHARACTERISTICS

The FRP composite materials used here have 70% of glass, aramid or carbon fibers, 5 to 25 μm in diameters, bound by a polyester resin (matrix) (figure 3) and accordingly are called GFRP, AFRP or CFRP. Their mechanical characteristics depend on the matrix and fibers (figure 4), and the tensile strength in the fiber direction is far higher than that of steel (figure 5). Behavior of FRP material at tensioning is linearly elastic until failure.

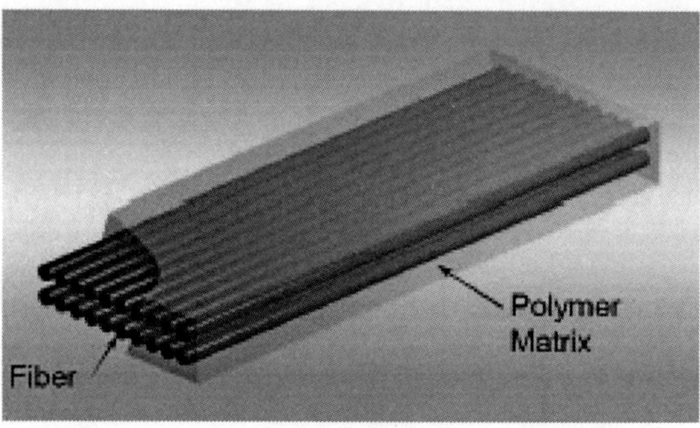


**Figure 3.** Composition of FRP.

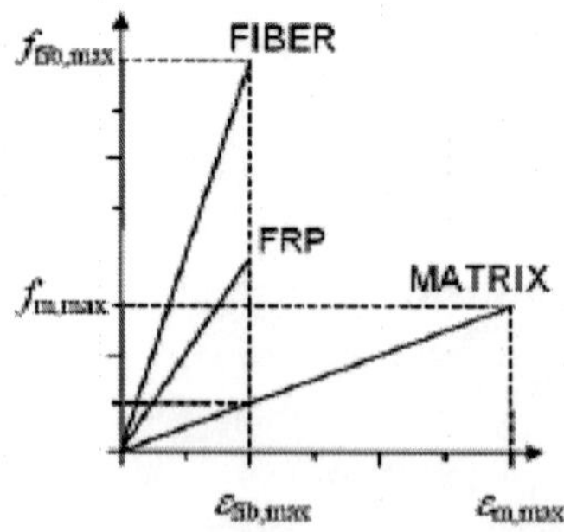


**Figure 4.** Component ratio.

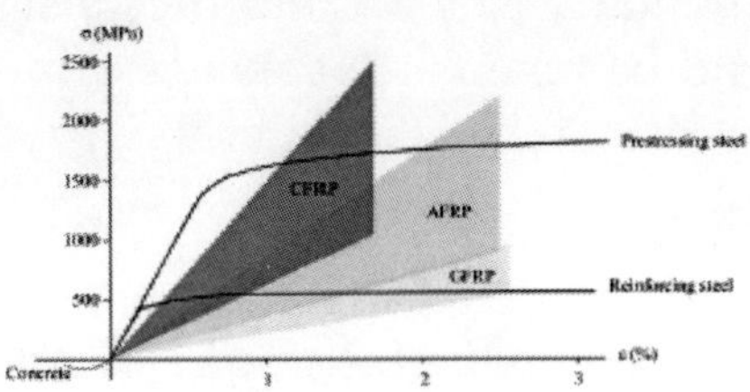


**Figure 5.** Tensile strength.

The FRP bars can be manufactured in an almost unlimited number of sub variants, that is, shapes. Thus, NSM FRP reinforcement can have square, rectangular or oval cross-section of the bar. The term "bars" is adopted as universal for all kinds of cross-section forms, while the term "strips" remains reserved for thin and narrow bands, strips [2]. Various forms of cross-section provide a variety of advantages and offer various possibilities for practical application. For instance, the square bars increase the specific rein- for cement area of the bar, whereas the circular bar cross sections are easier to anchor during pre-stressing. Narrow thin strips increase the surface coefficient and thus reduce the risk of adhesion loss, but require a thicker concrete coating for a given surface area of the cross section. The FRP bars are produced with a variety of surface textures, which have a considerable effect on the increase of adhesiveness of the NSM reinforcement. Their surface can be smooth or rough, with spiral grooves or ribbed. In practical application, the choice depends more on the actual conditions, such as the available thickness of concrete cover, availability of cost of certain types of FRP reinforcement.

Adhesives constitute a very important part of the FRP reinforcement systems which are most frequently on the basis of an epoxy resin or rarely cement, and they are used as a binder with concrete. The total bearing capacity of the reinforcement system considerably depends on their performance.

On the basis of opinions on application of FRP elements available in the literature, the following conclusions of their characteristics are drawn: 1) under a short-term load, they behave linearly- elastic until failure, 2) they are not ductile and thus limit the ductility to the reinforced elements 3) the value of the adhesion should be determined experimentally ( by a "pull-off" test), 4) the service life of the material should be taken into consideration [7].

## NSM FRP METHOD

The NSM method is based on the technique, whereby the bar-shaped or strip-shaped FRP elements are placed as additional reinforcement in the groove made in concrete cover, and they are embedded in epoxy or cement resins (adhesives), which creates adhesion with the concrete, and provides anchoring [2,12] (figure 6).

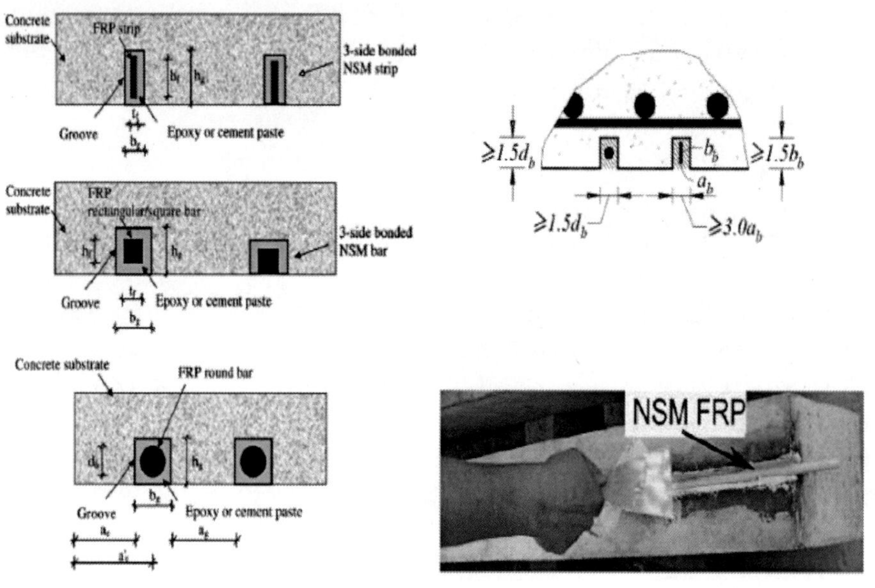


**Figure 6.**Application of NSM FRP "bar" elements.

Usage of the NSM reinforcement method via the FRP elements is justified in the fol- lowing cases: 1) if the reinforced surface is susceptible to damage; 2) if the reinforced surface is rough; 3) if the concrete surface has insufficient tensile strength, while the re- maiming part of the cross section does have the sufficient strength and 4) if the space available to install other types of reinforcement is insufficient. These are, in fact, the most common cases occurring in practice. The limitation in application are related to provision of sufficient thickness of the of concrete cover, which should be 1.5 times larger than the diameter of the used reinforcement bars (Ø6÷Ø16 mm). Available experience also sug- gests that the form of the FRP element, size and form of the, size and shape of the groove, types of fiber, types of adhesive, concrete tensile strength and concrete surface preparation methods should be taken into account. [3,4]. According to the up-to-date experience in application of the NSM systems, three kinds of failure are possible: 1) separation from the adhesive; 2) separation of the concrete; and 3) under tensile forces, failure of the FRP elements is possible.

By analyzing the behavior of various kinds of reinforcements reveal that the NSM method has a number of advantages in comparison to EBR FRP method because it is safer and exhibits higher durability in respect to the EBR [6]. Namely, this system of rein- for cement allows better adhesion, that is, anchoring, because the bars can be shaped (textured/ribbed) in the fabrication process, which provides a larger specific area for binding with concrete [11]. The NSM reinforcement can be more easily pre-stressed. Be- sides, thermal protection becomes easier, ant his is one of the basic issues in application of FRP reinforcement, due to the thermal instability of adhesives. Durability, which may become compromised due to the aggressive action of the environment, [8] or due to the physical damage is also enhanced when applying the NSM system, which makes this technology particularly suitable for reinforcement of the regions in beams and slabs where negative momentums occur.

## REZULTATS OF NSM GFRP REINFORCED BEAMS

At the Faculty of Civil Engineering and Architecture of Niš experimental testing of bearing capacity of beam girders reinforced by the GFRP reinforcement by loading until failure was performed. In figure 7, dimensions, reinforcement details and load application method are presented.

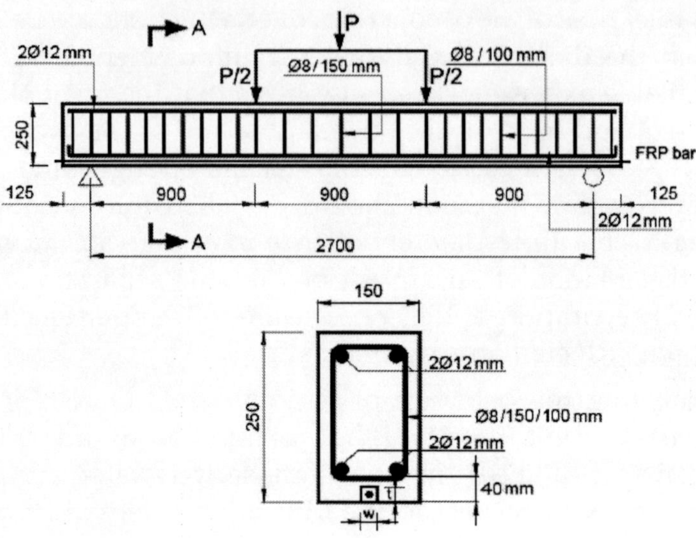


**Figure 7.** Details of reinforcement and load application method.

In Figure 8 a presentation of a beam with the GFRP reinforcement is given, and its de- formed shape under the maximum load. The load is applied with two concentrated forces at the thirds of the span (four point load). For reinforcement, the FRP bars G-rod Ø10 mm were used and the epoxy adhesive MapeWrap 11, by the Italian manufacturer MAPEI [5]. Recording of measuring data was performed by the acquisition system MGCplus, and the quasi-dynamic by reading of instruments for one second.

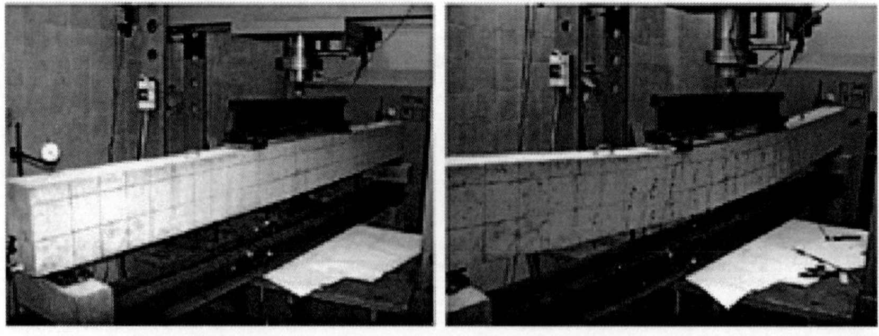

**Figure 8.** Experimental setup - prior to loading and under loading.

Experimental results presenting the relation load-deflection at the half of the span for a beam reinforced by a GFRP bar and for the control beam are given in the figure 9. It can be observed the diagram that the maximum load achieved by reinforcement was for 73% higher, that is, that by applying the NSM method, the maximum force was increased from 45 kN to 78 kN. Also notable is the satisfactory ductility of the reinforced beam which makes this method advantageous for seismic reinforcement. Until onset of the first cracks there is no difference in rigidity of the beams, it is generated in the part of the diagram from the first cracks to the onset of yielding in the reinforcement, and particularly after that point. The failure occurred due to separation of the epoxy resin and concrete due to tensile overstressing of concrete.

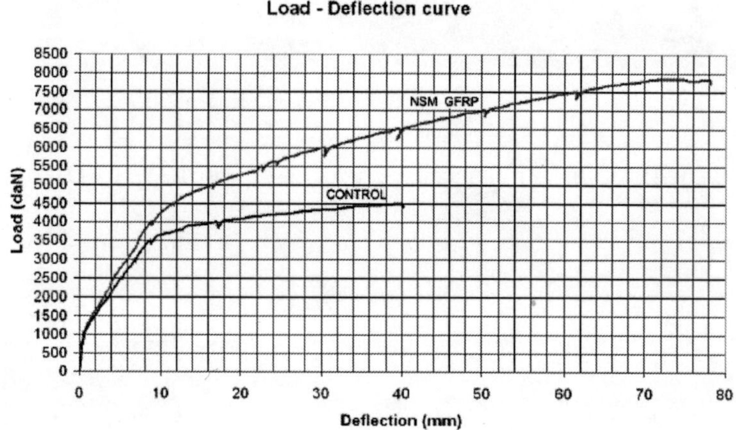


**Figure 9.**Diagrams load-deflection in L/2 of the reinforced and control beam.

Cracks distribution and their propagation have been measured during loading at 5 kN, which as a consequence has a certain decrease of force, particularly at high loading levels. For this reason, the curve is not ideally smooth. At the half of the span, the strain meter was installed whose LVDT measuring instrument is located at 20 mm from the upper and lower edge of the beam re- spectively. The strain meter in the tensed zone (with the gauge length of 200 mm) continuously measured the cracks width, that is, the sum of cracks widths along 200 mm of gauge length, and the results of local deformations are presented in the figure 11.

**Figure 10.**Distribution of measuring instruments on L/2 (LVDT and strain meter).

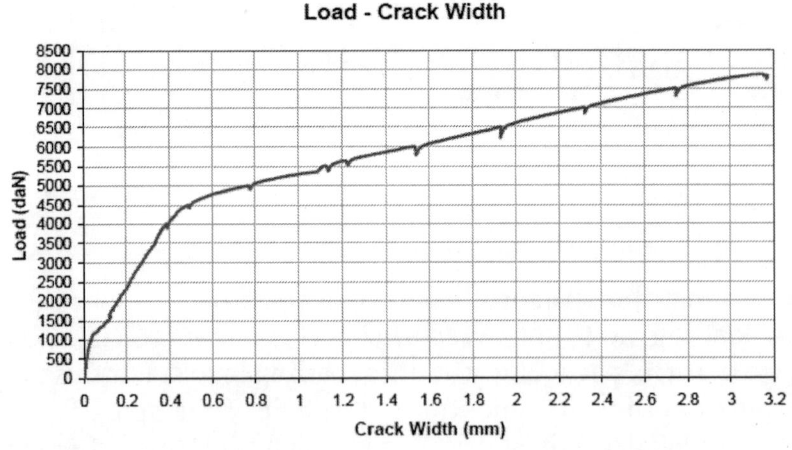


**Figure 11.**Diagram loading - crack width reinforced by the GFRP reinforcement.

From the diagram in figure 11, the sum of crack widths along the length of 200 mm in function of the loading can be analyzed. The same form of diagram, as in deflection can be observed, that is, occurrence of three characteristic zones: 1) until onset of the first cracks (0.04 mm), 2) since onset of the first cracks until the occurrence of yield in steel reinforcement (0.5) and 3) since the onset of the reinforcement yield until failure (3.15 mm). The same instrument measured elongation on the concrete surface at the level of the GFRP reinforcement, i.e. strain (figure 12). Through application of this procedure, it is possible to directly determine tensile strength of concrete at bending. Namely, via the value of strain obtained with the onset of the first crack, the tensile stress in concrete can be defined, which will be the subject of another paper.

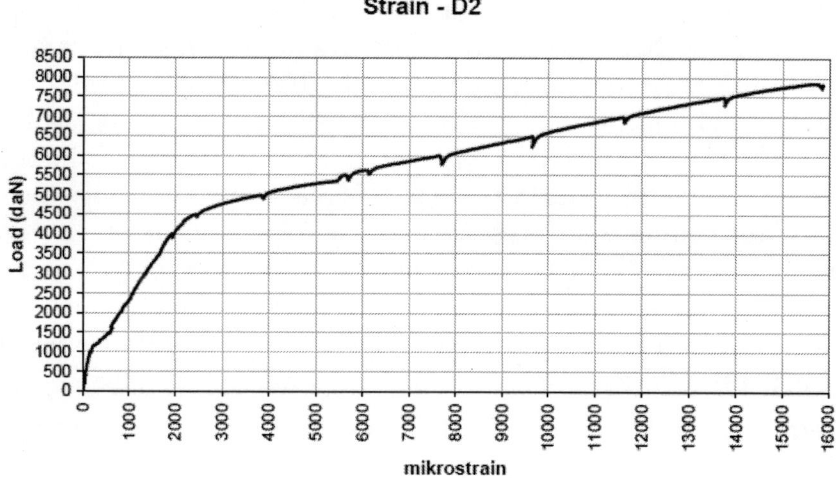


**Figure 12.**Tension strain diagram on the concrete surface.

From the diagram in figure 12 it may be observed that the first cracks occur at dilatation of approximately 200 με, that is, elongation of about 0.04 mm on the gauge length of 200 mm. Al- most the same values were obtained on the control beam, while the elongations occurring at yield and failure are lower than in the beam reinforced by the GFRP reinforcement.

## DISCUSSION OF RESULTS AND CONCLUSION

It is progressively evident that composite materials in the field of construction engineering are becoming the materials of the future. Price difference, which is a principal limiting parameter preventing wide application, is reducing each passing year, and the development of technology enhances the quality of the materials. It is the reason that the FRP reinforcement is becoming a very likely alternative to the conventional reinforcement method of addition of missing steel reinforcement. Application of the NSM rein- for cement method, as a relatively new one, offers great potential in restorations and strengthening of concrete structures and extension of their service life.

Analysis of the data from the available literature and independent research conducted in 2009 at the Faculty of Civil Engineering and Architecture of Nis, exhibited considerable increase of bearing capacity of the tested beams reinforced by the NSM method. In the actual case, when only one (additional) GFRP Ø10 mm is applied, a 73% higher maximum load was registered. In the higher loading phases and after monitoring of failure mechanisms, a significant ductility of beams reinforced by the GFRP reinforcement was manifested. The failure occurred at the joint of epoxy paste and concrete. In the situations when the rigidity is not the limiting parameter the GFRP rein- for cement should be favored over the CFRP bars because its cost is considerably lower.

Numerous advantages of the NSM FRP method of internal reinforcement are in many cases superior to the externally-applied FRP laminates. With a more complete regulations, and information of designs and easier accessibility of the FRP materials in the market, its wider application in practice can be expected.

This paper has been a result of the research projects no. 16001 and 16018 financed by the Ministry of Science of the Republic of Serbia.

We would like to thank the MAPEI company (the representatives in Belgrade) for the assistance in FRP material, used in experimental research.

## REFERENCES


1. ACI committee 440, "Guide for the Design and Construction of Concrete Reinforced with FRP Bars," ACI 440.1R-03, American Concrete Institute, Farmington Hills, Michigan, 2003, 41 pp.
2. ACI 440R-07, Report on Fiber-Reinforced Polymer (FRP) Reinforcement for Concrete Structures, Re- ported by ACI Committee 440, 2007.
3. Barros J., Dias S. Fortes A.,: Near surface mounted technique for the flexural and shear strengthening of concrete beams.
4. Concrete Society (CSC), Technical Report No.55: Design Guidance for Strengthening Concrete Struc- tures Using Fibre Composite Materials, 2004
5. Mapei FRP System, www.mapei.com
6. Fédération Internationale du Béton (FIB):Tecnical Report Bulletin 14: Externally Bonded FRP Reinforcement for RC Structures, Lausanne, 2001.
7. Folić, R., Glavardanov, D.: Analiza metoda pojačavanja armiranobetonskih elemenata lepljenjem vlaknastih kompozita (FRP), Izgradnja br. 5-6, 2006, str. 113-126.
8. Folić, R.: Durability design of concrete structures-Part 1: Analysis fundamentals, FACTA UNIVERSITATIS, Series A&CE, Vol. 7, No 1, 2009, pp. 1-18
9. Glavardanov, D. Folić, R.. Pojačavanje betonskih konstrukcija FRP elementima NSM sistemom. Materi- jali i konstrukcije, br. 4 2007., str. 29-35
10. Ranković S., Folić R. Mijalković. M.: Ojačanje AB greda FRP atmaturom postavljenom unutar zaštit- nog sloja betona, Zbornik radova GAF Niš, br. 23, decembar 2008., (st.39-47).

# Chapter 4

# POLYURETHANE STRUCTURAL ADHESIVES APPLIED IN AUTOMOTIVE COMPOSITE JOINTS

Josue Garcia Quinil,[II, *]; Gerson Marinucci[II]

[I]Masterpol Adhesives Technology, Rua Luiz Vaz de Camoes, 98, CEP 07210-007, São Paulo, SP, Brazil

[II]Nuclear and Energetic Research Institute - IPEN/CNEN-SP, Av. Prof. Lineu Prestes, 2242, CEP 05598-900, São Paulo, SP, Brazil

## ABSTRACT

In recent years structural adhesives technology has demonstrated great potential for application due to its capacity to transform complex structures into solid unitary and monolithic assemblies using different materials. Thus, seams or joints integrate these structures providing, besides a reduction in weight, a considerable increase in the mechanical resistance and stiffness. The increase in the industrial use of structural adhesives is mainly due to their ability to efficiently bond different materials in an irreversible manner, even replacing systems involving mechanical joints. In the automobile industry structural adhesives have been widely used for the bonding of metal substrates, thermoplastics and composites, frequently employing these in combination, particularly glass fiber and polyester resin composites molded using RTM and SMC processes. However, the use of urethane structural adhesives in applications involving composites and thermoplastics has been the subject of few investigations. In this study the effects of temperature and time on the shear strength of RTM, SMC and ABS joints, applying temperatures of -40, 25, 80, 120 and 177 °C and times of 20 minutes and 500 hours, were determined. The objective was to evaluate the performance under extreme conditions of use in order to assess whether these joints could be used in passenger or off-road vehicles. The results showed that the urethane structural adhesive promoted the efficient bonding of these materials, considering that due to the high adhesive strength the failures occurred in the substrates without adversely affecting the bonded area. For each test condition the joint failure modes were also determined.

## INTRODUCTION

The demands of modern industries such as those of the automotive, aeronautic and shipbuilding sectors, where there is a strong commitment to increasing productivity, with requirements for high quality indices, have lead to an ever increasing use of structural adhesives on their assembly lines. The advantages offered by structural adhesives in relation to traditional mechanical joints, such as welds, rivets or screws, include the possibility to bond distinct materials with different thermal expansion coefficients, obtaining monolithic structures which are mechanically extremely resistant [1]. It is also possible to achieve a reduction in weight, greater stiffness and better surface finish than bonds formed by mechanical fixing[2].

In recent decades, the automotive industry, in particular, has evaluated new materials aimed at attaining better vehicle performance and structural adhesives have contributed to this objective, permitting not only the bonding of traditional metal materials with polymeric materials and composites, but also the bonding between these types of materials themselves.

The bonding of parts using structural adhesives offers significant benefits in relation to traditional systems. The adhesive distributes the loads and stresses acting on the total bond area instead of concentrating them, allowing not only a uniform distribution of static and dynamic loads but also reducing the production and maintenance costs in relation to mechanically fixed systems. Furthermore, it ensures better electrical insulation, a reduction in corrosion and also a reduction in the vibration levels of assemblies with screws and rivets. Industrially, in many cases the methods of adhesive application offer higher productivity in the assembly processes[3]. Another important advantage resulting from an effective adhesion is a good seal between the bonded parts, inhibiting the passage of fluids through the joint and dispensing with the need for additional impermeabilization[4].

Several factors can affect the performance of a joint bonded by a structural adhesive, one of these being a variation in temperature[5-8]. These factors have been the focus of several studies reported in the literature, since they allow the identification and prevention of the conditions under which joints may fail and which hinder the good performance of the structure.

In the literature related to this subject there is a concentration of studies involving epoxy adhesives. Taib et al.[9]studied the strength of joints of polyester composites with glass fiber using epoxy adhesive, while Balkova et al. reported results for the shearing of specimens of pultruded substrates of reinforced polyester with glass fiber, bonded with epoxy adhesive, after exposure to a temperature of 60 °C[10]. The work of Kim et al.[7], also involving epoxy adhesive, the results showed the effect of the surface treatment of epoxy composites reinforced with carbon fibers under different temperature conditions and the modes of substrate rupture as a function of the surface treatment were evaluated.

The adhesive joint design must be selected considering the nature of its future application. Silva et al.[5]performed a study to evaluate joints designed with epoxy and bismaleimide adhesives at different temperatures. The joints were manufactured from composites, aluminum and titanium and aimed at aerospace applications. The authors evaluated the distribution of stresses along the joints by way of finite elements analysis.

Malucellia et al.[11] investigated the shear strength of joints bonded with a monocomponent polyurethane adhesive (base polyester) on substrates of polyoxypropylene, polypropylene and aluminum. The researchers investigated surface treatment methods, kinetics reaction and the thermal behavior of the adhesive and obtained as a result an increase in shear strength after treatment and the speed of the adhesive cure reaction by varying the air humidity and the parameters of the temperature of use. Thus, the aim of the present study was to investigate and evaluate two components urethane structural adhesives using substrates of composite and thermoplastic materials and also to present results of testing conditions which have not been previously extensively evaluated.

## FAILURE MODES CHARACTERISTIC OF ADHESIVE JOINTS

In order to identify the failure modes in shear tests, the guidelines of the ASTM D5573 [12] standard were followed, which classifies the rupture modes of adhesive joints. The main rupture modes suggested by the standard are shown in Figure I.

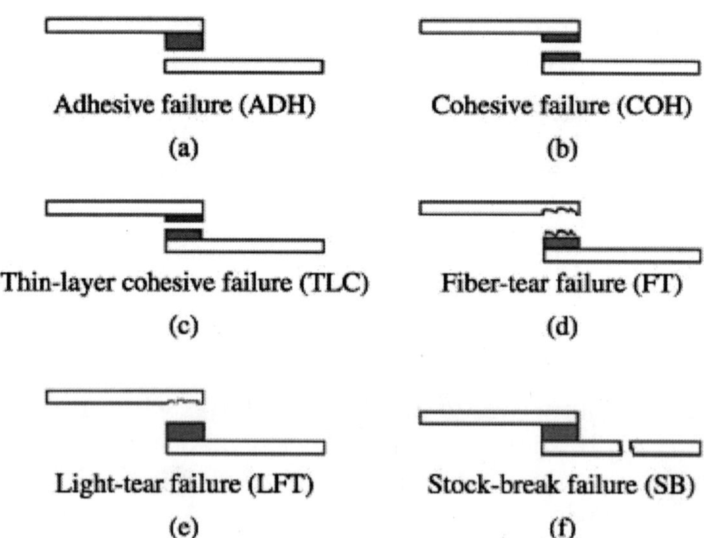


**Figure 1.** Representation of failure modes characteristic of adhesive joints submitted to shearing.

The factor which determines the failure mode of a specimen is the difference between the cohesive and adhesive resistances and the resistance of the substrate[13]. The failure will firstly occur at the point of least resistance.

The failure of the substrate, in turn, can occur in two ways: stock-break or fiber-tear. The previous concept also applies in these cases, that is, the failure will firstly occur through the path of least resistance Stock-break occurs when the specimen submitted to the shear test breaks outside the bonded area, while in the fiber-tear mode the failure arises when the specimen resists the applied stress but undergoes flexion during the test which can initiate the rupture process.

The failure of anisotropic substrates can occur in both ways. The composites used in this study have this quasi-isotropic character. Defects in the composite substrate such as air bubbles, differences in the polymeric matrix and reinforcement fractions and inefficient impregnation of the fibers favor the fiber-tear process, initiating a rupture process in the region close to the interface with the adhesive. Composite substrates which have a low number of defects tend to undergo stock-break failure.

It should be highlighted that in the analysis of the test results, not only the shear strength values should be considered but also the failure mode, given that in the substrates of composite materials this can occur via both fiber-tear and stock-break modes.

## MATERIALS AND METHODS

Masterpur Estrutural 300 urethane adhesive manufactured by Urepol-Masterpol Adesivos was used and it was applied with a pneumatic mixing dosing device.

The thermoplastic and composite substrates were produced at an automotive plant. The objective of this study was to investigate the capacity of the urethane structural adhesive to replace the use of mechanical joints with adhesive joints and to demonstrate that the adhesive is as or more efficient than the mechanical joint. ABS is the generic name for a family of amorphous thermoplastics formed by a combination of acrylonitrile monomers, butadiene and styrene which have a high dimensional stability, good surface appearance, easy processability, good electrical insulation and good chemical resistance. It is commonly employed in the manufacturing of parts for buses, for example, fenders, roof, fuel tank protector, and internal door panels, and is molded by the vacuum forming technique.

For the composite specimens laminated polyester reinforced with glass fiber was used, molded using RTM (Resin Transfer Molding) and SMC (Sheet Molding Compound) as used in parts for trucks, cars and buses.

The evaluation of the performance of the adhesive was carried out through the determination of the shear strength using the standard ASTM D3163. Five specimens were tested in each test for each of the proposed conditions and the results represent the average of the measurements and its respective expanded uncertainty. A total of around one hundred specimens were used.

The preparation of the surface of the composite and thermoplastic substrates was carried out by abrasion to remove the surface layer which could interfere in the adherence of the urethane adhesive, followed by cleaning with isopropyl alcohol.

The tests were conducted 24 hours after the bonding at different temperatures which may occur during the use of the joint in automotive parts. The temperatures applied were based on the norms WSBM11P27B of Ford and TMS 6900 of International Trucks, both restricted use. For the conditioning of the specimens at these temperatures an oven with air circulation and a cold chamber were used.

## RESULTS AND DISCUSSION

Firstly, the results for the shear strength at 25 °C are presented, which serves as a parameter for comparison with the other tests. The results obtained for the specimens at different temperatures are then reported. Lastly, the specimens were exposed to high temperatures followed by stabilization and then the test at ambient temperature.

### Shear strength of joints bonded at room temperature

The specimens were tested to evaluate the shear strength at 25 °C using a Kratos universal testing machine with a load cell of 1000 kgf, adopting a test velocity of 12.7 mm/min. The results together with their respective uncertainties are given in Table 1.

**Table 1.** Evaluation of shear strength of joints at 25 °C.

| Substrate | Shear strength (MPa) | Expanded uncertainty (MPa) | Failure mode |
|---|---|---|---|
| SMC | 5.8 | 0.4 | Fiber-tear |
| RTM | 6.1 | 0.3 | Stock-break |
| ABS | 3.6 | 0.1 | Stock-break |

Under the conditions of the test in Table 1 the failures occurred in the substrates while the adhesive remained undamaged. In the SMC there was fiber-tear failure and in the RTM and ABS stock-break failure, as shown inFigures 2-4, respectively. For the ABS specimens the shear strength was lower than the values obtained for the SMC and RTM, which can be interpreted as a natural occurrence considering that ABS has a lower tensile strength than the composites. This indicates that the bonding with the adhesive was efficient since the failures occurred in the substrate, maintain the integrity of the adhesive.

(a) (b)

**Figure 2.** Test specimens of SMC in adhesive shear test: (a) before the test, (b) after test where the arrow indicates the region of fiber-tear.

(a) (b)

**Figure 3.** Test specimens of RTM in adhesive shear test: (a) before the test, (b) after test where the arrows indicate the stock-break.

(a) (b)

**Figure 4.** Test specimens of ABS in adhesive shear test. (a) before the test, (b) after a test where the arrows indicate stock-break.

The adherence observed for the urethane adhesive is due to the efficient interaction between the adhesive and the substrate. In the case of the composite materials SMC and RTM, which traditionally have a greater percentage of polymeric matrix, it is possible to allow greater interaction between the adhesive and the polymeric matrix of the composite. This interaction can be explained by the high polarity of the polyester resin due to the presence mainly of hydroxyl radicals in the polymeric chain[14], which can interact with the urethane adhesive both electrostatically and through chemical reaction with the isocyanate groups present in the adhesive.

It can be observed in Figure 2 that the rupture of the SMC occurred through fiber-tear and the arrow in the figure indicates the region of the visible fibers. During the test it was observed that the specimens on receiving the mechanical loading underwent flexion leading to the beginning of the failure of the polymeric matrix followed by the rupture of the glass fibers. After the separation of the two parts of the specimen it can be verified that the glass fibers also underwent rupture and became exposed.

In the RTM, according to Figure 3, stock-break failure occurred, in contrast to the fiber-tear mode observed for the SMC.

The SMC substrate underwent flexion and consequently fiber-tear, whereas the RTM substrate showed only slight flexion and this was not sufficient to initiate failure by fiber-tear and instead stock-break failure occurred. The results obtained for the shear strength were 5.8 ± 0.4 MPa for SMC and 6.1 ± 0.3 MPa for RTM.

In the tests carried out with the ABS substrate the failure mode was similar to that observed for the RTM substrate, as shown in Figure 4. It can be observed that the fracture was transversal to the loading direction, presenting a flat form with an aspect characteristic of a pure traction test.

ABS is comprised of three monomers: acrylonitrile (A), butadiene (B) and styrene (S)[15]. Butadiene and styrene are monomers formed of carbon and hydrogen only and the conditions of application are not suitable for their efficient interaction with the urethane adhesive. The acrylonitrile group ($H_2C{=}CH{-}C{\equiv}N$), during the polymerization reacts through the breaking of the C=C double bond, the C≡N group then being available for interaction with the adhesive. This group can form bonds with the hydrogen atoms of the polyol hydroxyl groups of the urethane adhesive.

## Effect of temperature and time on the shear strength

For the design of an adhesive joint, besides the properties at ambient temperature it is important to also evaluate the performance under the conditions to which the joint could be submitted after being integrated into a structure. Variations in the temperature and relative air humidity and exposure to corrosive chemical substances, associated with time, are agents which can create conditions which reduce the integrity of the substrate and the adhesive, reducing the load capacity of the joint.

It is recommended that these external conditions are evaluated in isolation and also in a sequence called a cycle, where variations present in the environment where the joint will be used can be simulated. In this study the effects of temperature and time on the behavior of joints bonded with urethane adhesive were investigated.

Since the main use of urethane structural adhesives is in the automotive industry, the external conditions employed in this study were based on the same norms WSBM11P27B and TMS 6900.

### *Effect of temperature*

It is known that the use of polymeric materials is restricted when they are submitted to harsh temperature conditions, and maximum temperature limits are almost always established, for instance, 170 or 200 °C. The determination of the conditions at cryogenic temperatures is rare.

In order to determine the shear strength as a function of temperature the tests were carried out under five temperature conditions: -40, 25,

80, 120 and 177 °C, after 24 hours of bonding. The test involving temperature is important since it allows the maximum capacity of the joint to bear mechanical loading at different temperatures to be established.

## SMC substrate

The determination of the shear strength as a function of temperature for the SMC substrates was carried out under the five conditions described above and the results can be found in Table 2. The shear strength at 25 °C represents an initial parameter for comparison with the other values and was established as a reference. The test performed at -40 °C led to a reduction in the joint strength of 3.4% when compared with that at 25 °C, with the occurrence of fiber-tear, however given the uncertainty of the measurement this was not considered to be significant.

**Table 2.** Shear strength of SMC specimens at different temperatures.

| Temperature (°C) | Shear strength (MPa) | Expanded uncertainty (MPa) | Variation (%) | Failure mode |
|---|---|---|---|---|
| –40 | 5.6 | 0.4 | –3.4 | Fiber-tear |
| 25 | 5.8 | 0.4 | --- | Fiber-tear |
| 80 | 5.2 | 0.3 | –10.3 | Fiber-tear |
| 120 | 1.9 | 0.1 | –67.2 | Fiber-tear |
| 177 | 0.5 | 0.1 | –91.4 | Cohesive |

In all of the tests conducted at above 25 °C a decrease in the shear strength occurred for all joints. At 80 °C there was a reduction in the strength of 10.3%, while at 120 °C the reduction was 67.2%, with failure occurring via the fiber-tear mode at both temperatures. This indicates that the laminated matrix of the SMC was probably affected by temperatures above 80 °C. At a temperature of 177 °C there was cohesive failure, with the joint showing a strength of 0.5 MPa, which represents a reduction of 91.4%. Under this condition the failure occurred soon after the beginning of the test which is why the strength was so low. If this temperature had not affected the adhesive, the rupture of the substrate would probably have occurred at a value lower that those observed at 120 °C.

## RTM SUBSTRATE

In a way analogous to that of SMC, a study was carried out using RTM substrates. The RTM fibers are arranged in the form of sheets, while the SMC material is comprised of a mixture of fibers cut with resin and mineral loads. The results obtained for the RTM joints at temperatures of -40, 25, 80, 120 and 177 °C can be observed in Table 3.

**Table 3.** Shear strength of RTM specimens at different temperatures.

| Temperature (°C) | Shear strength (MPa) | Expanded uncertainty (MPa) | Variation (%) | Failure mode |
|---|---|---|---|---|
| –40 | 5.9 | 0.3 | –3.3 | Stock-break |
| 25 | 6.1 | 0.3 | --- | Stock-break |
| 80 | 5.0 | 0.2 | –18.0 | Stock-break |
| 120 | 2.1 | 0.1 | –65.6 | Stock-break |
| 177 | 0.4 | 0.1 | –93.4 | Cohesive |

The test carried out at -40 °C showed a reduction of only 3.3% in the shear strength compared with that conducted at ambient temperature, and led to stock-break failure.

Once again, at all temperatures above 25 °C a decrease in the shear strength occurred. At 80 °C there was an 18% reduction while at 120 °C the reduction was greater, with a value of 65.6%. At both temperatures stock-break failure occurred, which indicates that the specimens were affected by the temperature, this being attributed to the polymeric matrix. The test performed at 177 °C led to cohesive failure, with a strength of only 0.4 MPa, which represents a significant reduction in relation to the ambient temperature. Since the failure was cohesive, the urethane adhesive was demonstrated to not be suitable for application at such high temperatures, as in the case of SMC.

## ABS substrate

The study on the ABS specimens was carried out under the same conditions applied to the SMC and RTM composites, except for the exposure to temperatures above 120 °C, since the substrate would not resist such conditions. The results obtained for the ABS joints at temperatures of -40, 25 and 80 °C are shown in Table 4.

**Table 4.** Shear strength of ABS specimens at different temperatures.

| Temperature (°C) | Shear strength (MPa) | Expanded Uncertainty (MPa) | Variation (%) | Failure Mode |
|---|---|---|---|---|
| –40 | 3.4 | 0.1 | –5.6 | Stock-break |
| 25 | 3.6 | 0.1 | --- | Stock-break |
| 80 | 3.6 | 0.1 | 0 | Stock-break |

The test carried out at -40 °C showed a reduction of 5.6%, which is higher than in the case of SMC and RTM, with stock-break failure occurring. At 80 °C there was no variation in the shear strength and the ABS showed the same failure mode as that observed at ambient temperature, indicating that at 80 °C the ABS did not undergo degradation.

## Effect of temperature and time of joint exposure

The shear strength of the bonded joints was evaluated after exposure to high temperatures for a period of time followed by stabilization at 25 °C and testing. Two conditions were evaluated according to the requirements of the norms WSBM11P27B and TMS 6900 for vehicles. The first was exposure to 90 °C for 500 hours (condition 1) and the second exposure to 177 °C for 20 minutes (condition 2). In the case of condition 1 the objective was to evaluate the effect of temperature for a long exposure time while in the case of condition 2 the aim was to evaluate the effect of a high temperature peak for a short period of time. For both conditions, according to the guidelines of the cited norms, at the end of the exposure to the respective temperatures for the set times the test was carried out at ambient temperature. Tables

5 and 6 show the shear strength results at a temperature of 25 °C after the exposures described above.

**Table 5.** Shear strength of SMC, RTM and ABS specimens subjected to condition 1 - exposure for 500 hours at 90 °C.

| Substrate | Shear Strength condition 1 (MPa) | Expanded Uncertainty (MPa) | Variation (%) | Failure Mode |
|---|---|---|---|---|
| SMC | 5.5 | 0.4 | –5.2 | Stock-break |
| RTM | 5.9 | 0.3 | –3.3 | Stock-break |
| ABS | 3.4 | 0.1 | –5.6 | Stock-break |

**Table 6.** Shear strength of SMC and RTM specimens subjected to condition 2-20 minutes exposure to 177 °C.

| Substrate | Shear Strength condition 2 (MPa) | Expanded Uncertainty (MPa) | Variation (%) | Failure Mode |
|---|---|---|---|---|
| SMC | 3.4 | | –41.4 | Fiber-tear |
| RTM | 3.2 | | –47.5 | Fiber-tear |

On comparing these results with those obtained under the initial condition of 25 °C (section 4.1), it can be noted that there was a slight reduction in the shear strength for condition 1 and the decrease was more accentuated for condition 2. These findings were attributed to the fragilization of the substrates.

In the case of the SMC which had a strength of 5.8 ± 0.4 MPa at 25 °C with a fiber-tear failure, condition 1 led to a 5.2% reduction and the failure mode was different, this time occurring via the stock-break mode. Applying condition 2 led to fiber-tear failure with a reduction in the shear strength of 41.4%. For the RTM, condition 1 led to a 3.3% reduction while condition 2 resulted in a 47.5% reduction in shear strength and the failure modes were, respectively, stock-break and fiber-tear. It can be observed that the effects of conditions 1 and 2 were very similar for the RTM and SMC both in terms of the reduction in strength and the percentage reduction. It was observed for the composite substrates that the greater the exposure temperature, even for a shorter time, the greater the effect of the strength reduction of the joint. The effect of a temperature of 177 °C with a short exposure time, only 20 minutes, was considerably more destructive than the long exposure (500 hours) to a lower temperature of 90 °C.

In the case of the ABS substrates it was not possible to evaluate the behavior at 177 °C due to its low melting temperature (105 °C). Thus, only condition 1 was applied and there was a reduction of 5.6% with failure occurring via the stock-break mode, a result close to that observed for the SMC substrate.

## CONCLUSIONS

The urethane adhesive showed an excellent performance in the bonding of composite and thermoplastic joints within the temperature range of -40 to 120 °C, showing a rupture of the substrate outside the bonding area. This shows that the adhesive was able to bear the mechanical force without undergoing failure. For a condition of 177 °C there was a considerable reduction in the shear strength with cohesive failure, indicating that the adhesive cannot be recommended for continuous use at this temperature.

The results showed that application of the conditions of 500 hours/90 °C and 20 minutes/177 °C led to a decrease in the shear strength due to failures in the substrates. This results from the effect of the exposure temperatures being above the Tg of the polymeric matrix of the composite substrates and also above the ABS deformation temperature, which leads to a breaking of the polymeric chains fragilizing the substrate. No alterations in the behavior of the adhesive was observed at these temperatures

## ACKNOWLEDGEMENTS

The authors acknowledge the technical support of the Masterpol Adhesives Technology and IPEN-Energetic and Nuclear Research Institute.

## REFERENCES

1. Baldan A. Adhesively-bonded joints in metallic alloys, polymers and composite materials: Mechanical and environmental durability performance. *Journal of Materials Science*. 2004; 39:4729-4797.http://dx.doi.org/10.1023/B:JMSC.0000035317.87118.
2. Hou TH, Boston KG, Baughman JM, Walker S and Johnston WM. Composite-to-composite Bonding using Scotch-Weld™ AF-555M Structural Adhesive. *Journal of Reinforced Plastics and Composites*. 2010; 29(11):1702-1711.http://dx.doi.org/10.1177/0731684409341679
3. Stewart R, Goodship V, Guild F, Green M and Farrow F. Investigation and demonstration of the durability of air plasma pre-treatment on polypropylene automotive bumpers. *International Journal of Adhesion and Adhesives*. 2005; 25:93-99. http://dx.doi.org/10.1016/j.ijadhadh.2004.04.001
4. Borsellino C, Calabrese L, Bella G and Valenza A. Comparisons of processing and strength properties of two adhesive systems for composite joints. *International Journal of Adhesion and Adhesives*. 2007; 27:446-457.http://dx.doi.org/10.1016/j.ijadhadh.2006.01.012
5. Silva LFM and Adams RD. Joint strength predictions for adhesive joints to be used over a wide temperature range. *International Journal of Adhesion and Adhesives*. 2007; 27:362-379.http://dx.doi.org/10.1016/j.ijadhadh.2006.09.007
6. Goeiju WC, Tooren VMJL and Beukers A. Composite adhesive joints under cyclic loading. *Materials & Design*. 1999; 20:213-221. http://dx.doi.org/10.1016/S0261-3069(99)00032-1
7. Kim JK and Lee DG. Characteristics of plasma surface treated composite adhesive joints at high environmental temperature. *Composite Structure*. 2002; 57:37-46. http://dx.doi.org/10.1016/S0263-8223(02)00060-0
8. Khalili SMR, Mittal RK and Kalibar SG. A study of the mechanical properties of steel/aluminium/GRP laminates.*Materials Science and Engineering*: A. 2005; 412:137-140.http://dx.doi.org/10.1016/j.msea.2005.08.016
9. Taib AA, Boukhili R, Achiou S, Gordon S and Boukehili H. Bonding composites. *International Journal of Adhesion and Adhesives*. 2006; 26:226-236.
10. Balkova R, Holcnerova S and Cech V. Testing of adhesives for bonding of polymer composites. *International Journal of Adhesion and Adhesives*. 2002; 22:291-295. http://dx.doi.org/10.1016/S0143-7496(02)00006-4

11. Malucelli G, Priolla A, Ferrero F, Quaglia A, Frigione M and Carfagna C. Polyurethane resin-based adhesives: curing reaction and properties of cured systems. *International Journal of Adhesion and Adhesives*. 2005; 25:87-91.
12. American Society for Testing and Materials - ASTM. *ASTM D5573*: Standard Practice for Classifying Failure Modes in Fiber-Reinforced-Plastic (FRP) Joints. ASTM; 2005.
13. Cognard P. *Handbook of adhesives and sealants*. Versailles: Elsevier; 2006. v. 2.
14. Sanchez EMS. Unsaturated polyester resins: Relation and properties. Influence of addition of poly (styrene-b-isoprene-b-styrene) resins on the mechanical properties. [Tese]. Campinas: Universidade Estadual de Campinas; 1996.
15. Brady GS, Clauser HR and Vaccari JA. *Materials Handbook*. 15th ed. New York: McGraw-Hill; 2002.

# Chapter 5

# APPLICATION OF MARTENSITIC SMA ALLOYS AS PASSIVE DAMPERS OF GFRP LAMINATED COMPOSITES

M. Bocciolone, M. Carnevale, A. Collina, N. Lecis, A. Lo Conte, B. Previtali

Politecnico di Milano, Department of Mechanical Engineering, Via La Masa, 1 Milano (Italy)

C.A. Biffi, P.Bassani, A. Tuissi

National Research Council CNR, Institute for Energetics and Interphases, Corso Promessi Sposi, 29 Lecco (Italy)

## ABSTRACT

This paper describes the application of SMA (Shape Memory Alloy) materials to enhance the passive damping of GFRP (Glass Fiber Reinforced Plastic) laminated composite. The SMA has been embedded as reinforcement in the GFRP laminated composite and a SMA/GFRP hybrid composite has been obtained. Two SMA alloys have been studied as reinforcement and characterized by thermo-mechanical tests. The architecture of the hybrid composite has been numerically optimized in order to enhance the structural damping of the host GFRP laminated, without significant changes of the specific weight and of the flexural stiffness. The design and the resultant high damping material are interesting and will be useful in general for applications related to passive damping. The application to a new designed lateral horn of railway collector of the Italian high speed trains is discussed.

## INTRODUCTION

Over the past few years, due to the increasing technological demand for the passive suppression of mechanical vibrations, there has been a growing interest in developing high damping materials. While a consistent amount of literature on the damping of SMAs (Shape Memory Alloys) has focused on the active damping of small-amplitude vibrations devoted either to applications of pseudo elasticity or during the super elastic phase, less attention has been paid to the passive damping behavior at small amplitude harmonic vibrations with varying frequency ranges. Small amplitude and passive vibration damping can be important in mechanical applications, where random broadband excitations are present in a typical operational environment.

Shape memory alloys are characterized by two phases: an austenitic phase, obtained at high temperatures and a martensitic one, obtained at low temperatures; the direct and reverse phase transformation between the two phases occurs in a certain temperature range, depending on the composition and heat treatment of the alloy [1]. High damping can be observed during direct and reverse transformations, and, consequently, during limited temperature intervals. Several types of shape memory alloys, such as Cu based SMAs, show high values of internal friction, even in the martensitic state [2-4]. The fairly high intrinsic damping of these martensitic phases has been associated with the atoms and defect motion and with the re-orientation of the marten site twin variants under stress [5]. From a practical point of view, this intrinsic energy dissipation mechanism of the martensitic phase over wider temperature ranges offers an interesting perspective for the application of SMA alloys in the martensitic state as passive dampers of low amplitude mechanical vibrations for automotive, aerospace and other dynamic applications [6, 7] The idea of adopting the damping properties of the martensitic state means that the service temperature has to be lower than the transformation temperature.

The high cost, the relatively high specific weight, and the relatively low specific modulus preclude the use of monolithic SMA as a bulk material in many structures; however, interesting examples of composites or hybrid composites, designed for damping, can be found in literature [8, 9]. Since the concept of SMA hybrid composites was first proposed in 1988 by Rogers et al. [10], these composites have attracted enormous attention in terms of improving creep and fatigue properties [11], strength [12] and damping capacity [13], as well as to control the shape or vibration response property [14]. Among these materials, the SMA/Glass-Fiber Reinforced Plastic (GFRP) hybrid composite - in which the embedded SMA is used to improve structural damping, thereby saving weight and stiffness, is especially important due to the wide potential or effective technological application of GFRP in all, real-life engineering environments [15, 16]. This paper describes a new architecture of SMA/GFRP hybrid composite numerically optimized in order to enhance the structural damping of the host GFRP laminated, without significant changes of the specific weight and of the flexural stiffness [7].

Two SMA alloys have been studied as reinforcement and characterized by termo-mechanical tests.

The design and the resultant high damping material are interesting and will be useful in general for applications related to passive damping, i.e. wind turbine blades or automotive components. Since the research began from studies about the dynamic optimization of the collector of the Italian high speed trains, this application is the guideline of the paper.

## DESIGN CONSIDERATIONS

The flexural modes of the collector of the high-speed train (Fig. 1) play a significant role in terms of its dynamic behavior and its lateral horns - usually made of fiber glass/epoxy resin laminated - are not able to guarantee the required level of structural damping. To optimize the dynamic behavior of the collector, structural damping must be enhanced, in order to avoid in-depth modification in terms of mass and stiffness [17, 18].

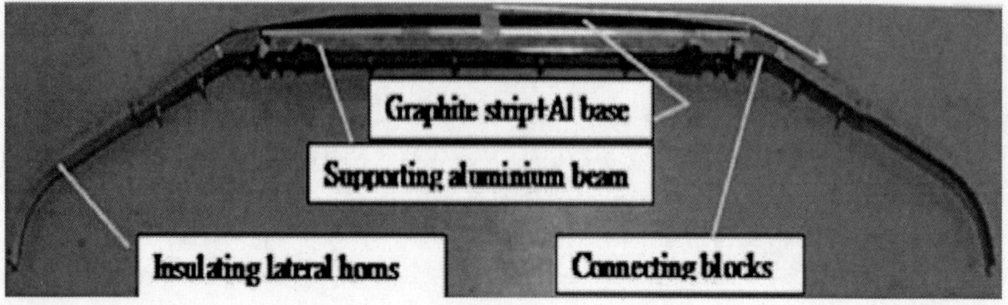


**Figure 1.** Italian high-speed train collector.

To enhance the damping of a Glass Fiber Reinforced Polymer (GFRP) beam and shell through passive vibration suppression, a shape memory alloy (SMA) can either be embedded as reinforcement in the GFRP part. Unlike GFRP, SMA alloys have a higher storage modulus and a higher specific damping [19, 20]. Due to its higher storage modulus and, assuming that the interface guarantees the right load transfer, the SMA material is capable of storing more specific elastic energy than the GFRP and, as consequence, is able to take maximum advantage of its higher specific damping in order to enhance the structural damping of the hybrid composite. Therefore the partial substitution of the GFRP material with SMA material in various shapes and forms is expected to significantly increase the damping properties of the hybrid composite.

In [21], thanks to the high damping of the NiTi in its martensitic the authors proposed the application of Ti-Ni SMA alloy wires as "smart fibers" embedded in this conventional GFRP material in order to create new horns with an enhanced flexural damping capacity related to the first flexural mode. Two series of 13 wires, with a diameter of 0.56 mm, were embedded below the upper and the lower surfaces of the horn as shown in Fig. 2.

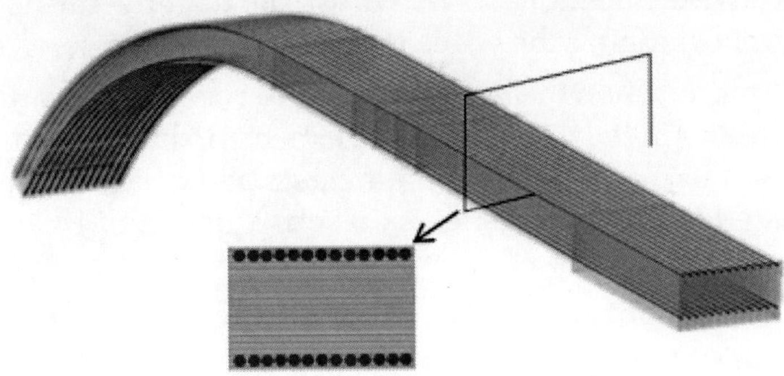

**Figure 2.** Lateral horn made from a GFRP laminated composite with two layers of embedded NiTi wires.

A first prototype of the horn was manufactured and then tested dynamically in a laboratory. Experimental results showed that not only do the NiTi wires collaborate in the composite but also that the non-dimensional damping of the first flexural mode increases by about 11% with respect to the non-dimensional damping of the original horn made only of GFRP. Conversely, the first natural frequency increases by about 5%. According to these results, it can be expected that the higher level of the vibration damping, required for real application of the proposed architecture of the horn, can be obtained with a proper volume fraction of the SMA wires; however, the conflicting consequences of the introduction of SMA wires between increasing of the structural damping and increasing of the first natural frequencies is addressed. For this reason, the same paper also discusses the implementation of a FE model of the horn in order to perform a dynamic analysis and to calculate the related non-dimensional damping. The results are presented for three configurations: the horns in their original configuration (only GRFP), the horns made from GRRP with two series of 13 embedded wires, as in the real prototype shown in Fig. 3, and the horn made from GRRP with two series of 20 embedded wires. As expected, by increasing the number of wires, the vibration damping of the horn increases; however, the undesirable side effect of the NiTi wires is that the flexural frequency increases by 10%, thus indicating that the requirements of the renewed design cannot be satisfied. Additionally, some consideration should also be given to the architecture of the lateral horn with embedded SMA wires. The manufacturing process involved in positioning the wires within the matrix is very complex.

Furthermore, when the horn is cooled to room temperature, high residual stresses in the SMA wires are expected, due to the mismatch between the thermal expansion coefficients of the SMA material and the laminated GFRP composite. As a consequence, SMA fiber pull-out can easily occur [22].

The information generated and the experience gained in our previous paper will be used to propose a new SMA/GFRP hybrid composite, based on a synergistic contribution of the parameters associated with the SMA material such as specific damping, specific stiffness and volume fraction as well as those associated with the host, including flexural rigidity, SMA through-the-thickness location and SMA-host interfacial strength. Particular attention has been paid to manufacturability and the cost-effectiveness of the component made with the proposed new composite.

## THE ARCHITECTURE OF THE NEW HYBRID COMPOSITE

The architecture of the new proposed hybrid composite is shown in Fig. 3. The base composite is a symmetric angle-ply laminated of fiber glass/epoxy resin (3M-SP250 S29A). The elastic properties of the unidirectional layer are shown in Tab. 1. The sequence of lamination is [45/-45]15. The elastic modulus (E11=E22) of the laminated fiber glass/epoxy resin with this sequence of lamination is 17 GPa. The loss factor, obtained by performing DMA tests (Q800 DMA, TA Instruments) on a 35x11.36x1.29 mm3 sample with a lamination sequence of [45/-45]3 at room temperature (indicatively 20 °C), is 0.8 % and, in the range (1-70 Hz) results not dependent on the frequency value. Two SMA sheets are embedded below the upper and the lower surface of the layered GFRP. Moreover, patterning of the SMA sheets is introduced to improve adhesion between the SMA sheets and GFRP layers, to avoid the delamination of the composite and to maximize the mechanical energy transmission to the SMA element. The geometry of the patterning is optimized with respect to the improvement in terms of the composite damping and the adhesion at SMA/glass fiber interface [15].

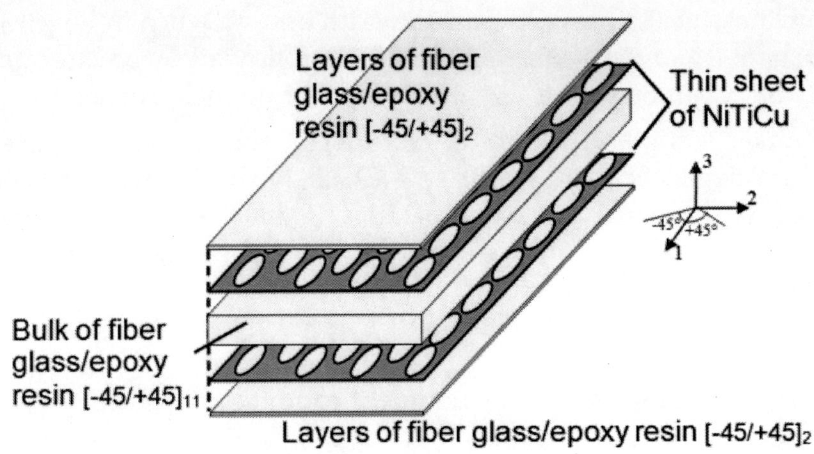


**Figure 3.** Architecture of the new proposed hybrid composite.

**Table 1.** Elastic properties of the fiber glass/epoxy resin 3M-SP250 S29A.

| Traction | | Compression | | | |
|---|---|---|---|---|---|
| $E_{xx}$ [GPa] | $E_{yy}$ [GPa] | $E_{xx}$ [GPa] | $E_{yy}$ [GPa] | $G_{xy}$ [GPa] | $n_{xy}$ |
| 45.7 | 13.5 | 47.8 | 12.8 | 5.4 | 0.27 |

## SMA MATERIALS

A NiTiCu and CuZnAl alloys are proposed for the reinfocement. NiTiCu is a well-known system, obtained by partial substitution of Ni with copper, whose damping capacity is high and stable versus the vibration time and aging effect of martensite. CuZnAl alloy has the highest damping capacity of all high damping metals [3] as well as a relatively high and appropriate modulus of elasticity. The NiTiCu and CuZnAl alloys were prepared by means of a vacuum induction melting system. The nominal atomic composition of the SMAs produced are Ni40Ti50Cu10 and Cu66Zn24Al10. The ingots were hot forged and hot and cold rolled to achieve sheets boasting a thickness of 0.3 mm (30-mm in width and 400-mm in length). The final heat treatments were 450°C for 1h, followed by a water quench and 750 °C for 30 min followed by a water quench, respectively.

Small samples, weighing about 15 mg, were subjected to Differential Scanning Calorimetry (DSC), with a heating and cooling rate of 10 °C/min in the range 10-110 °C (Q100 DSC, TA Instruments). The thermographs shown in Fig. 4, outline the characteristic transformation temperatures: Mf=32 °C; Ms=49 °C; As=52 °C and, Af=61 °C for the $Ni_{40}Ti_{50}Cu_{10}$ and Mf=50 °C; Ms=63 °C; As=60 °C and, Af=68 °C for the $Cu_{66}Zn_{24}Al_{10}$. The DSC scan confirms, for both materials, the martensitic structure at room temperature.

The damping properties of the alloy were evaluated by means of dynamic mechanical analyses (Q800 DMA, TA Instruments) carried out on thin sheet (12x1x0.15 mm3) of examined materials. Different values of the frequency and strain amplitude were analyzed. The specimen was subjected to a sinusoidal strain ($\varepsilon$) with amplitude of 0.01%, 0.05%, and 0.1% and to a oscillation frequency (f) of 2, 10, 20, and 35 Hz. The phase lag ($\delta$) between the applied stress and the resultant strain were measured in the temperature range [-20 °C÷120 °C] with a cooling/heating rate of 1 °C/min. As a result, the intrinsic damping (tan $\delta$ of the SMA was obtained as a function of the temperature.

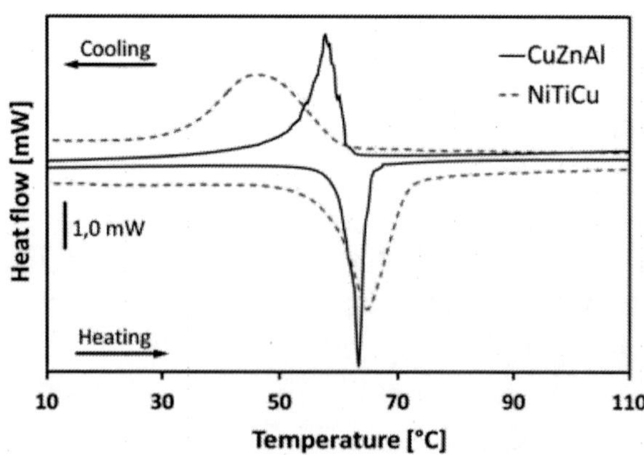


**Figure 4.** DSC scan of the investigated materials, heating/cooling rate: 10 °C/min.

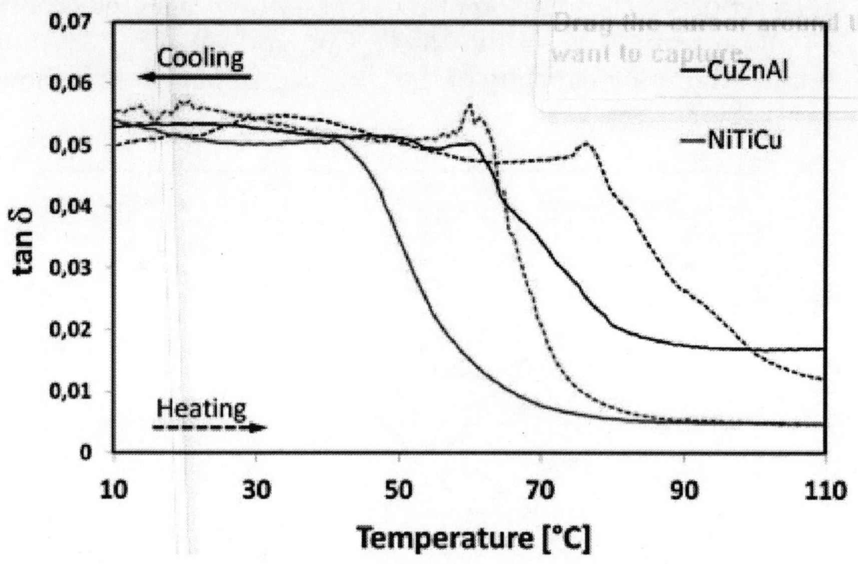


**Figure 5.** tan δ vs. temperature. DMA analysis: heating rate 1°C/min, frequency 10 Hz, strain amplitude 0.05%.

Fig. 5 shows the intrinsic damping tan δ as a function of temperature for an 0.05% strain amplitude, at a 10 Hz frequency. The high damping capacity of the marten site phase can be observed at temperatures of below the As temperature, while, at higher temperatures, the parent phase (austenite phase) exhibits low tan δ values. No significant influence of frequency was observed in the range (2÷35 Hz) tested.

Owing to the fact that only a few grains were present in the cross section of the DMA specimens, results were validated through cyclic stress strain tests on larger samples.

A MTS quasi-static testing machine with a load cell of 10 kN is used for cyclic tension tests, at room temperature, on specimens measuring 200x20 $mm^2$. The cycle frequency is fixed as equal to 0.05 Hz, based on the assumption that the SMA damping properties do not vary significantly at low frequencies. A series of 16 tests was performed by controlling the cyclic elongation using a 50 mm gage length extensometer. The deformation amplitude ranges between 2 $10^4$ and 1 $10^3$, while the minimum value is about 3 $10^4$. In each test, 10 cycles were run at the specified deformation amplitude.

Fig. 6 reports a comparison of the stress-strain hysteresis loop at 0.075% strain amplitude for the $Ni_{40}Ti_{50}Cu_{10}$ and $Cu_{66}Zn_{24}Al_{10}$ alloys. The highest damping capacity of the $Cu_{66}Zn_{24}Al_{10}$ are clearly shown.

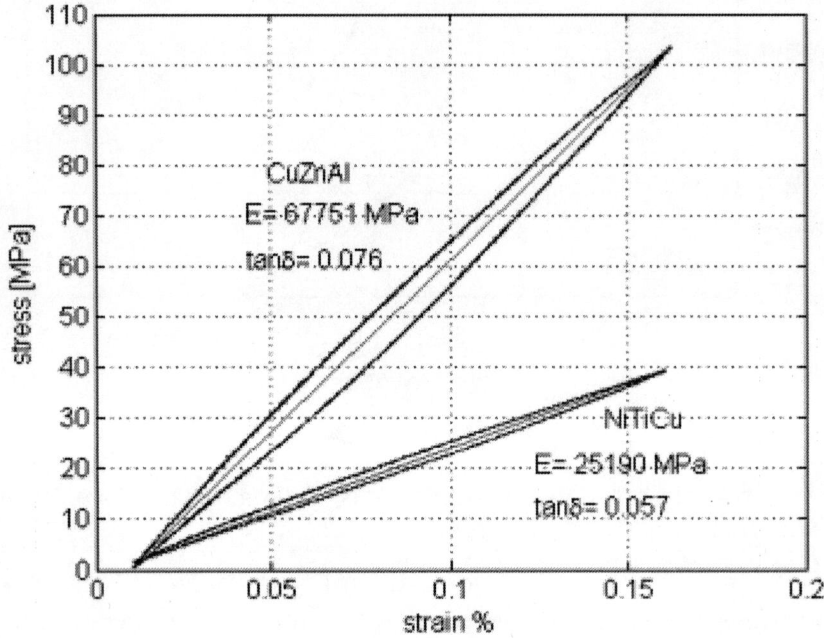


**Figure 7.** Stress-strain hysteresis loop for CuZnAl and NiTiCu during MTS tensile test at 0.075% strain amplitude.

The following parameters have been calculated at each strain amplitudes:

1) tan $\delta_i$ value, obtained from the single cycle as:

$$\tan \delta_i = \frac{\Delta W_i}{2\pi W_i}$$

where $\Delta W_i$ is the energy dissipated in the ith cycle of oscillation for unit volume, representing the area enclosed within the stress–strain hysteresis loop, and Wi is the maximum stored energy in the same cycle of oscillation per unit volume;

2) storage modulus $E_i$, evaluated as $\Delta\sigma/\Delta\varepsilon$ from the $i^{th}$ hysteresis loop;

3) SMA specific damping capacity tan δ and SMA storage modulus, obtained as the average values over the ten stress–strain loops performed.

The results are shown in Figs. 7 and 8 both for $Ni_{40}Ti_{50}Cu_{10}$ and $Cu_{66}Zn_{24}Al_{10}$. The loss factor shows an almost linear increase of the damping as a function of the strain amplitude. For $Cu_{66}Zn_{24}Al_{10}$ a nonlinear increase at about 0.06% of strain was observed. The tan d discontinuity at 0.05% strain amplitude was also observed during DMA tests. This confirms that this particular trend of tan $\delta$ is associated with a certain behavior of the Cu based SMA. In order to understand this phenomenon better, accurate experiments will be performed in future developments. By comparison of results of Fig. 7 with DMA results of Fig. 5 at a frequency of 10 Hz, a good agreement can be observed. As shown by the results of Fig. 8, the storage modulus decreases as the strain amplitude increases.

## NUMERICAL ANALYSES

Finite Element Analysis was used in conjunction with a Modal Strain Energy approach to study the change in the natural frequencies and the loss factor of a beam or shell, made from angle-ply laminated fiber glass/epoxy resin ($[45/-45]_{15}$), when two thin sheets of the examined SMA alloy are embedded in the hybrid composite based on the architecture of Fig. 3. The hybrid composite is orthotropic with 1, 2, 3 as axes of orthotropy. The Finite Element Model of the hybrid composite is a solid, beam-shaped model. The fiber glass/epoxy resin base composite (200x20x4.6 $mm^3$) is modeled using 20-node brick elements with reduced integration; the thin SMA sheets are modeled with eight node shell elements. The constraint between the upper and lower surface of the GFRP laminated composite and the thin SMA sheet is a tie constraint. The beam is clamped at one end (see Fig. 9).

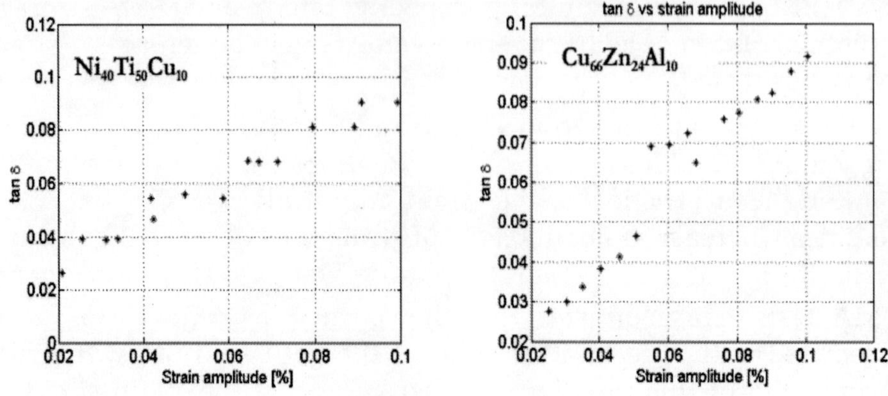


**Figure 7.** tan δ vs. strain amplitude. Tensile test: f=10 Hz, room temperature.

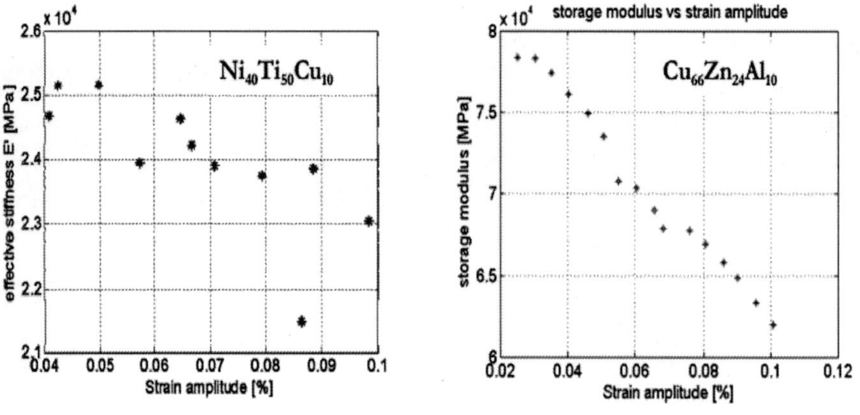


**Figure 8.** Storage modulus vs. strain amplitude. Tensile test: f=10 Hz, room temperature.

A linear elastic, undamped, modal finite element analysis was performed using the Abaqus 6.7 code to extract the first natural frequency of the beam. The fiber glass/epoxy resin laminated composite has been assumed as an homogenous material having an appropriate density value and a linear elastic behavior with an elastic modulus equal to the elastic modulus $E=E_{11}$ of the GFRP base composite, and a Poisson coefficient equal to $v=v_{12}$, of the angle-ply laminated fiber glass/epoxy resin (see Tab. 2). The behavior of the

SMA material has also been assumed as being linear elastic with the properties shown in Tab. 2.

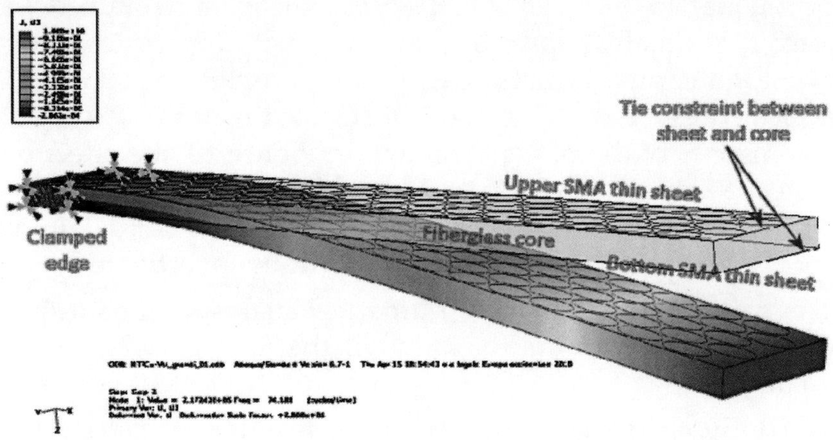


**Figure 9.** FEM model of the hybrid composite and first flexural deformed shape with superimpose the vertical displacement when the displacement of the free end is 1 mm

**Table 2.** Input data for the FEM analyses

| Input data | NiTiCu | CuZnAl | Glass fiber |
|---|---|---|---|
| Young Modulus E [MPa] | 23000 | 78000 | 17000 |
| Poisson Coefficient [-] | 0.30 | 0.20 | 0.35 |
| Density [kg/mm$^3$] | 6.50 $10^{-6}$ | 7.40 $10^{-6}$ | 1.86 $10^{-6}$ |

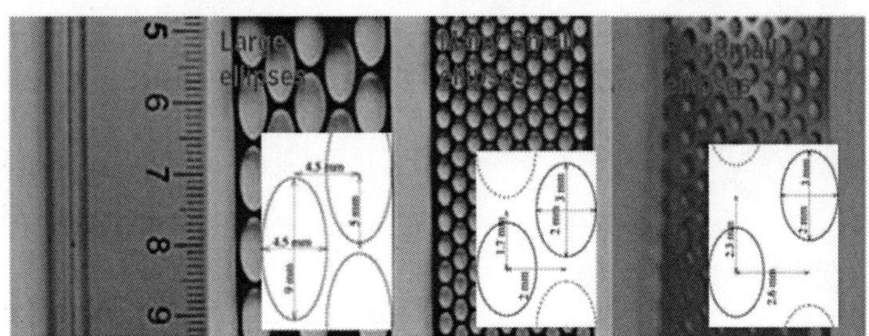

**Figure 10.** Thin sheets following laser micro-cutting with three different patterns, having a different SMA surface/total surface ratio.

Three different patterns of thin SMA sheets, characterized by the same elliptical shape, but having a different hole surface/SMA surface ratio are proposed (Fig. 10). The hole surface/SMA surface ratio is equal to 2.4, 2.25 and 0.65 for the pattern named Large ellipses, Few Small Ellipses and Many Small Ellipses respectively. Likewise, for comparison, the case of thin, un-patterned SMA sheets were also considered. Given the two selected material of Tab. 2 and the geometries of the patterns shown in Figure 10, the effect of the thin sheet thickness on the first natural frequency of the beam and on the splitting of the modal strain energy between GFRP laminated composite and reinforcement was numerically investigated.

The preliminar result of the numerical analyses is about the first natural frequency. As the thickness of the fiber glass/epoxy resin base composite is kept constant (4.6 mm), if the thicknesses (t) of the SMA equal to 0.1, 0.2 and 0.3 mm, the specific weight ($\delta$) of the hybrid composite is slightly affected by the presence of the SMA insert, and the increase of the first natural frequency with thin, patterned SMA-Large ellipse, SMA-Many Small ellipse, and SMA-Few Small ellipse, sheets is limited and almost independent on the thickness. Only for the unpatterned SMA the first natural frequency is thickness dependent and the maximum variation is about 5% when the thickness is 0.3 mm. Results are reported in Tab. 3 for the first vibrational mode of the hybrid composite with, SMA-Large ellipses, SMA- Many Small ellipses and SMA-not patterned, $Ni_{40}Ti_{50}Cu_{10}$ thin sheets embedded.

**Table 3.** FEM analysis estimated loss factor of the hybrid composite with Ni40Ti50Cu10 thin sheets embedded

| | t [mm] | f [Hz] | $a_{thin\ sheets}$ [%] | $a_{GFRP\ laminated\ composite}$ [%] | tan δ [%] |
|---|---|---|---|---|---|
| *Fiber glass/ epoxy resin* | -- | 71.1 | -- | 100 | 0.84 |
| SMA -Large ellipse | 0.1 | 73.9 | 4.9 | 95.1 | 0.92 |
| | 0.2 | 74.1 | 8.5 | 91.5 | 1.10 |
| | 0.3 | 74.2 | 11.6 | 88.4 | 1.30 |
| SMA- Small ellipse | 0.1 | 74.1 | 6.2 | 93.7 | 1.04 |
| | 0.2 | 74.3 | 10.7 | 89.3 | 1.18 |
| | 0.3 | 74.3 | 14.4 | 85.6 | 1.30 |
| *SMA-not patterned* | 0.1 | 75.5 | 15.5 | 84.5 | 1.76 |
| | 0.2 | 76.6 | 26.7 | 73.2 | 1.68 |
| | 0.3 | 77.5 | 35.3 | 64.6 | 1.94 |

The second aspect investigated by numerical analyses is the percentage of modal strain energy stored in the thin SMA sheets ( $\alpha$ thin sheet) and in the GFRP laminated composite ($\alpha$ GFRP laminated), for the first flexural mode, that is directly associated to the calculation of the loss factor of the hybrid composite (tan $\delta$). The Modal Strain Energy approach gives us [23]:

$$\tan\delta = \alpha_{thinsheets} \tan\delta_{thinsheets} + \alpha_{GFRPlaminatedcomposite} \tan\delta_{GFRPlaminatedcomposite} \quad (1)$$

where

$\alpha$ thin sheet and $\alpha$ GFRP laminated are respectively the percentage of the modal strain energy stored in the thin sheets and in the GFRP laminated composite;

tan $\delta$ thin sheet and tan $\delta$ GFRP laminated are the loss factors of the corresponding materials.

The loss factor of the hybrid composite was calculated according to Eq. (1) are also reported in Tab. 3. As can be observed in the sixth column of Tab. 3, when the $Ni_{40}Ti_{50}Cu_{10}$ patterned sheets are embedded in the composite, the loss factor is significantly enhanced. The composites with unpatterned inserts perform very well in terms of damping, but are likely to delaminate during service life [24]. As opposed to the fiber glass/epoxy resin beam, the composites with thin, inserted patterned sheets show a significant improvement in terms of damping capacity, provided that the thickness of the SMA layer is greater than 0.2 mm. Thanks to the presence of hollow ellipses, a significant improvement in the insert/GFRP adhesion is also expected. The beam having Many Small ellipses, patterned SMA sheets gives a better performance in terms of damping.

As regards the thickness effect, it is necessary to point out that the composite loss factor increases with the thickness of the thin SMA sheets. On the other hand, technological aspects related to the laser micro-cutting process have to be considered when the sheet thickness used in the layered composite is increased. Laser cutting becomes more difficult when the thickness is considerably increased owing to the fact that the process speed has to be reduced and a multi pass strategy adopted. Both actions decrease cutting edge quality in terms of dross, roughness and thermal damage. Since finishing operations, targeted at removing the layer affected by the laser beam are intentionally disregarded because they increase process costs and time, a thickness of 0.2 mm seems to be a good compromise between damping and workability. A comparison between the damping performance of the hybrid composite with $Ni_{40}Ti_{50}Cu_{10}$ and $Cu_{66}Zn_{24}Al_{10}$ thin sheets, is reported in Tab. 4.

$Cu_{66}Zn_{24}Al_{10}$ thanks to its higher storage modulus stores up to 25 % of the elastic energy of the deformed beam with respect to the 10% of the modal strain energy stored by the $Ni_{40}Ti_{50}Cu_{10}$ reinforcement with the same Many Small Ellipses patterns. As consequence the higher specific damping of the $Cu_{66}Zn_{24}Al_{10}$ alloy participate with a greater contribution to the enhancement of the structural damping of the hybrid composite

Moreover a modification of the pattern, from the SMA_Many Small Ellipses to the SMA_Few Small Ellipses, with a hole surface/ SMA surface ratio is equal to 0.65 (see Fig. 11), allows a further significant increase of the modal strain energy stored in the SMA thin sheets and of the final structural damping of the composite.

**Table 4.** FEM analysis estimated loss factor of the hybrid composite, comparison between Ni40Ti50Cu10 and Cu66Zn24Al10 thin sheets embedded

| | t [mm] | f [Hz] | $\alpha_{thin\ sheets}$ [%] | $\alpha_{GFRP\ laminated\ composite}$ [%] | tan δ [%] |
|---|---|---|---|---|---|
| $Ni_{40}Ti_{50}Cu_{10}$ – Many Small ellipse | 0.2 | 74.3 | 10.7 | 89.3 | 1.18 |
| $Cu_{66}Zn_{24}Al_{10}$ – Many Small ellipse | 0.2 | 66.1 | 25 | 75 | 1.60 |
| $Cu_{66}Zn_{24}Al_{10}$ – Few Small ellipse | 0.2 | 69 | 37 | 69 | 2.04 |

These numerical results have been validated by experimental decay tests performed on sample of the proposed new composite. The experimental tests are described in detail in [13, 25].

## SMA/GFRP LATERAL HORN

The new architecture of a SMA/GFRP hybrid composite profile is proposed for the design of an SMA/GFRP lateral horn.
In this case the host composite is a symmetric angle-ply laminate of fiber glass/epoxy resin with a stacking sequence of $(45/-45)_n$, where n varies, thus accounting for the variation in the horn thickness, ranging from 49 (thicker section) to 24 at the opposite end. On the basis of the results of the previous paragraph a $Cu_{66}Zn_{24}Al_{10}$ material and the Few Small Ellipses pattern has been selected for this application. As regards the thickness effect, it is necessary to point out that the composite loss factor increases with the thickness of the SMA sheets. Even if laser micro-cutting becomes more difficult when the thickness increases, a thickness of 0.3 mm has been adopted in order to take the maximum advantage from the passive damping of the SMA reinforcements.

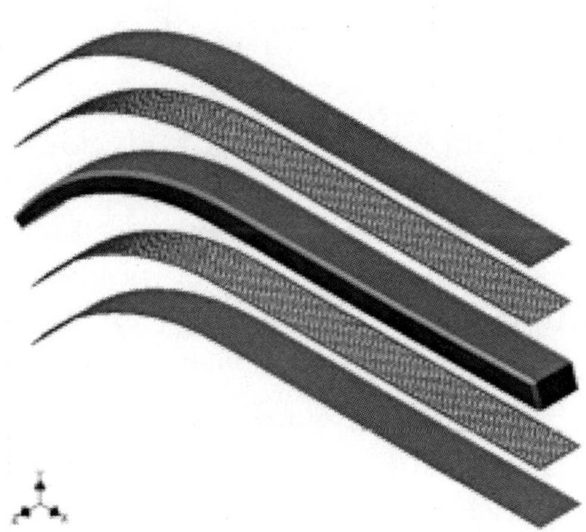

**Figure 11.** Architecture of the SMA//GFRP hybrid composite horn. Grey $[-45/+45]_n$ layered GFRP host composite. Brown: Embedded CuZnAl SMA thin patterned sheets.

A prototype of the horn has been made. Firstly, the regular pattern of elliptical holes has been produced by means of laser cutting using a pulsed nanosecond fiber laser (IPG Photonics YLP50). The process parameters used to perform the micro-cutting process of the elliptical pattern are shown in Tab. 5.

The maximum average power available, together with the highest pulse frequency and process speed were chosen to ensure through cutting and reduced thermal damage. A double laser pass strategy was required in order to guarantee stable cutting through the entire thickness of the SMA sheet [26].

**Table 5.** Main process parameters used in laser micro-cutting of the elliptical pattern

| Process speed [mm/s] | Average power [W] | Pulse frequency [kHz] | Shielding gas |
|---|---|---|---|
| 5 | 50 | 80 | Argon at 5 bar |

In the second step, the laminated composite is assembled. The horn mould is filled first with four layers $[+45/-45]_4$ of unidirectional fiber glass/epoxy resin and covered with the first patterned thin SMA sheet; the sequence of lamination $[+45/-45]_n$ is added and covered by the second thin SMA sheet and by an additional four layers $[+45/-45]_4$ of unidirectional fiber glass/epoxy resin. As previously reported, parameter n is dependent on the thickness of the cross section along the axis of the horn. Any air bubbles are eliminated by means of a vacuum bag and the hybrid composite is then cured in an autoclave.

In order to study the dynamic properties of the proposed hybrid layered architecture and to evaluate the improvement of the damping capacity due to the introduction of the SMA insert, three samples of the horn were produced. The first one was manufactured as described, while the second one was produced using the same architecture as the first one. However, in this case, commercial brass sheets were used in the place of the SMA sheets. The third one was only made with GFRP. Manufacture of an extra horn with a commercial brass insert is targeted at evaluating/excluding any possible contribution of a metal sheet/host composite interface to the damping of the SMA hybrid composite horn. As shown in Fig. 12, following the cure cycle in an autoclave, the three horn samples are completed.

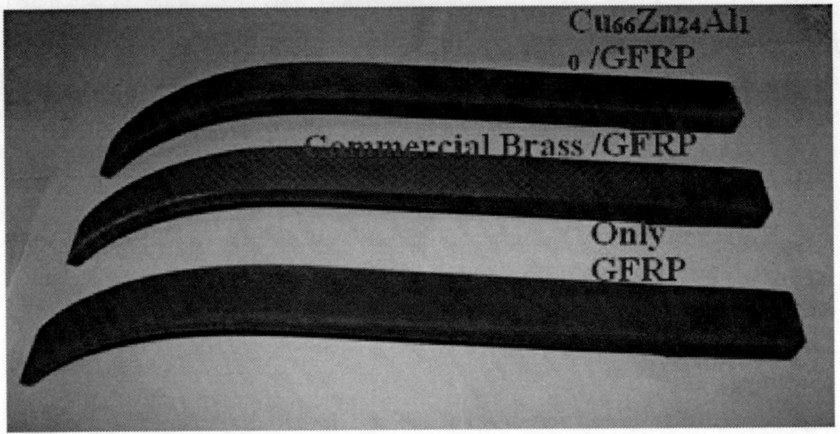


**Figure 12.** Final appearance of the three horn samples.

## EXPERIMENTAL TESTS

The non-dimensional damping, related to the first flexural mode of the three horns manufactured, was experimentally measured by performing a series of decay tests with the horn in a single cantilever configuration. During the tests, the end of the horn, designed to be connected to the structure of the collector, was clamped on a steel fixture, while the other end was loaded with an initial vertical displacement and then released to oscillate freely [22]. The transient response was recorded in terms of the vertical displacement of the section of the horn at a distance of 150 mm from the clamp. The displacement was measured by means of a lase-triangulation sensor (MEL M5L/10).

The non-dimensional damping was evaluated as follows:

$$h_n = \frac{\delta_n}{2\pi} \qquad \text{where} \qquad \delta_n = \ln\left(\frac{x_n}{x_{n+1}}\right) \tag{2}$$

where $\delta_n$ is the logarithmic attenuation coefficient, $x_n$ is the vertical displacement amplitude of the horn at 150 mm from the clamping at the nth oscillation of the transient response and $x^{n+1}$ is the same displacement amplitude at the (n+1)$^{th}$ oscillation. During transient decay, the amplitude of each successive oscillation decreases, meaning that Eq. (2) allows us to obtain the dependence of the non-dimensional damping from the displacement amplitude.

In order to compare the non-dimensional damping (h) with the loss factor (tan $\delta$), assumed as an index of the intrinsic damping in section Design considerations, the following consideration can be made. The loss factor (tan $\delta$) of the oscillatory beam, considered as a single degree of freedom system, is defined as the ratio

$$\tan\delta = \frac{E_d}{2\pi U} \qquad (3)$$

where $E_d$ is the energy loss per cycle and U is the strain energy.

Assuming that the motion is entirely due to the first flexural mode

$$U = \frac{1}{2} k_1 q_1^2 \qquad (4)$$

$$E_d = \pi \cdot c_1 \omega_1 q_1^2 \qquad (5)$$

where $k_1$ is the modal stiffness, $q_1$ is the modal coordinate, $c_1$ is the modal damping, and $\omega_1$ is the frequency of the first flexural mode of the beam.

On account of $k_1 = m_1\omega_1^2$ where $m_1$ is the modal mass of the beam and, from the definition of non-dimensional damping $c_1 = h_1(2m_1\omega_1)$, Eq. (3) can be rewritten as:

$$\tan\delta = \frac{\pi \cdot c_1 \omega_1 q_1^2}{2\pi \frac{1}{2} k_1 q_1^2} = \frac{h_1(2m_1\omega_1)\omega_1 q_1^2}{\pi m_1 \omega_1^2 q_1^2} = 2h \qquad (6)$$

Eq. (6) recalls that the loss factor is equal to twice the non-dimensional damping value. The natural frequency of vibration of the first flexural mode can be identified by Fourier transform of the waveform of the transient response of the horns acquired during the decay tests reported in Fig. 13. In this figure, the transient response of the horn with a Commercial brass patterned insert is actually superimposed on the only GFRP horn and, for sake of clarity, is not reported. The first natural frequencies of the three horn samples are shown in Tab. 6 and are not affected by the integration of the SMA/Commercial brass. The total weight of the horns is also reported in the same table and is only slightly affected by the integration of the SMA/Commercial brass. As will be shown below, the reduced thickness of the SMA/Commercial brass sheets and the patterns with a ratio between the surface of the holes and the total surface of the SMA sheet, equal to 0.65, not only allows us to maintain the structural and dynamic behavior of the component but also to improve the damping capacity. The beneficial effect of the integration of an SMA insert in the host GFRP horn on its damping capacity is apparent in Fig. 13.

The horn manufactured as an SMA/GFRP hybrid composite has a larger logarithm attenuation coefficient of the first flexural mode both as regards the horn made only from GFRP as well as the one manufactured as a Commercial brass/GFRP hybrid composite. As previously mentioned in Fig. 13, the transient response of the latter two is practically superimposed.

The results of experimental non-dimensional damping, associated with the first flexural mode, calculated from the logarithmic decrement of Fig. 13 when the displacement is equal to 1mm, are shown in Tab. 7.

The comparison between the non-dimensional damping of the Only GFRP horn and the non-dimensional damping of the horn with commercial Brass reinforcement allows us to exclude any significant contribution of the metal sheet/host composite interface to the damping of the SMA hybrid composite horn. When the SMA patterned sheets is embedded in the composite, due to its high specific damping the non-dimensional damping of the horn is significantly enhanced.

**Table 6.** First natural frequency and total weight of the three horn samples

| | First natural frequency [Hz] | Total weight of the horn [g] |
|---|---|---|
| Only GFRP | 75.5 | 168 |
| Commercial brass/GFRP | 75.3 | 182 |
| SMA/ GFRP | 74.8 | 180 |

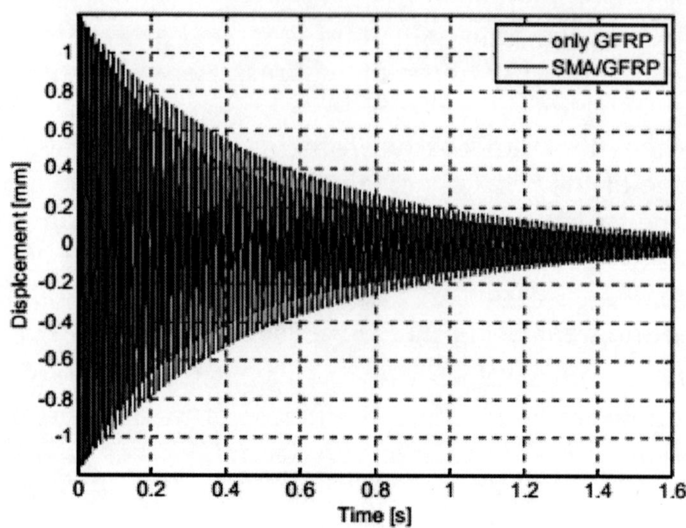


**Figure 13.** Transient of the displacement in the section at 150 mm from the clamped section, for the horn with a SMA patterned insert and for the horn made only from GFRP.

Table 7: Tan $\delta$ e non-dimensional damping (h) for the three horns when the displacement is equal to 1 mm.

| Horn | Tan δ | h |
|---|---|---|
| Only GFRP | 0.8 | 0.4 |
| Commercial Brass | 0.84 | 0.42 |
| $Cu_{66}Zn_{24}Al_{10}$/GFRP | 1.4 | 0.7 |

## CONCLUSIONS

To enforce the application of SMA alloys for passive damping in mechanical systems, shape memory alloy-based composites with thin, embedded, patterned SMA sheets were numerically optimized and fabricated. It has be shown that the new architecture of the hybrid composite is applicable to improve the structural damping of the host GFRP laminated, without significant changes of the specific weight and of the flexural stiffness. A salient result of this study is that the proposed hybrid composite architecture can be used for passive control of GFRP slender structure. Patches of patterned SMA thin sheets can be embedded only in the area of highest deformation of the GFRP structure according to the considered modal shape, to improve their passive damping capacity without any design modification and significant effects on the dynamic behavior.

The pattern of the thin SMA sheets proved to be a key feature in the optimization process of the flexural stiffness, weight and damping capacity of the hybrid composite. At the same time, this feature plays a positive role in the improvement of adhesion, and in the load transfer, between the GFRP laminated composite and the SMA reinforcements. The application of the proposed hybrid composite for a new design of a GFRP lateral horn of a railway pantograph has been proposed. A first prototype of the horn was manufactured. The salient points of the manufacturing process included the production of a CuZnAl SMA alloy having an appropriate transformation temperature; the laser micro-cutting of the CuZnAl SMA sheets and the fabrication of the component without any purpose-made alignment device.

## ACKNOWLEDGMENT

The authors would like to thank Mako Shark S.r.l–Viale Montecuccoli 18, 23843 Dolzago (LC), Italy, for manufacture of the laminated horns.

## REFERENCES

1. K. Otsuka, C.M. Wayman, Shape Memory Materials, Cambridge University Press, Cambridge, (1998).
2. A. Biscarini, B. Coluzzi, G. Mazzolai et al., J. Alloys Compd., 356 (2003) 669.
3. M.O. Moroni, R. Saldivia, M. Sarrazin, A. Sepulved, Material Science and Engineering, A335 (2002) 313.
4. Y.H. Li, S.W. Liu, H.C. Jiang, et al., Journal of Alloy and Coumponds, 430 (2007) 149.

5. J. Van Humbeeck, J. Alloys Compd., 355 (2003) 58.
6. J. Van Humbeek, S. Kustov, Smart Mater. Struct., 14 (2005) 171.
7. J. Raghavan, T. Bartkiewicz, S. Boyko, M. Kupriyanov, et al., Composites: Part B, 41 (2010) 214.
8. Ya Xu, K.Otsuka, H.Yoshida, H.Nagai, R.Oishi, H.Horikawa, T.Kishi, Intermetallics, 10 (2002) 361.
9. R. Zhang, Q. Q. Ni, A. Masuda, T. Yamamura, M. Iwamoto, Composite Structures, 74 (2006) 389.
10. Rogers CA, Liang C, Barker D. In: Proceedings of US Army Research Office Workshop on Smart Materials, Structures, and Mathematical Issues, Blacksburg, USA (1988) 39.
11. M. Dolce, D. Cardone, Int. J. of Mechanical Sciences, 47 (2005) 1693.
12. D.S. Li, X.P. Zhanga, Z.P. Xionga, Y.-W. Mai, Journal of Alloys and Compounds, 490 (2010) L15.
13. C.A. Biffi, P. Bassani, A. Tuissi, M. Carnevale, N. Lecis, A. Lo Conte, B Previtali, Functional Materials Letters, 5(1) (2012) 1250014.
14. G. Zhou, P. Lloyd, Composites Science and Technology, 69 (2009) 2034.
15. Y. Matsuzaki, T. Ikeda, C Boller, Smart Mater. Struct., 14 (2005) 343
16. F. Taheri-Behrooz, F Taheri, R. Hosseinzadeh, Material and Design, 32 (2011) 2023.
17. A. Collina, S. Melzi, In: AITC-AIT, Conference, Parma, Italy, (2006).
18. A. Collina, A. Lo Conte, M. Carnevale, Proc. IMechE Part F: J. Rail and Rapid Transit, 223 (2009) 1.
19. S.K. Wu, H.C. Lin, Journal of Alloys and Compounds, 355 (2003) 72.
20. C. Remillat, M. R. Hassan, F. Scarpa, Transactions of the ASME, 128, (2006), 260.
21. A. Tuissi, P. Bassani, A. Casati, M. Bocciolone A. Collina, M. Carnevale, A. Lo Conte, B. Previtali, JMEPEG, 18 (2009) 612.
22. K.T. Lau, A. W.Chan, S. Q. Shi, L. M. Zhou, Materials and Design, 23 (2002) 265.
23. G. Lepoittevin, G. Kress, Materials and Design, 31 (2010) 14.
24. Q. Q. Ni, R. Zang, T. Natsuki, M. Iwamoto, Composites Structures, 2007 (79) 501.
25. P. Bassani, C.A. Biffi, M. Carnevale, N. Lecis, B. Previtali, A. Lo Conte, Materials and Design, 45 (2013) 88.
26. B. Previtali, S. Arnaboldi, P. Bassani, et al., In: 10th Biennial Conference on Engineering Systems Design and Analysis, ESDA 2010, Istanbul (2010).

# Chapter 6

# FIBER REINFOCED POLYMER USED FOR FLOODING PROTECTION OF ENGINEERING STRUCTURES MADE OF RC AND BRICK MASONRY

Gabriel Oprisan*, Vlad Munteanu, Nicolae Taranu and Alina Lazar

Gheorghe Asachi" Technical University, Jassy, Department of Civil and Industrial Engineering.

## ABSTRACT

Urban and rural floods are becoming nowadays a frequent problem to be dealt with, by both the population and the authorities. Floods and flood related natural disasters act against the civil, industrial and agricultural structures by the hydrostatic and hydrodynamic pressures of water. A set of protective solutions based on Fiber Reinforced Polymer (FRP) composite materials, for structural elements of buildings subjected to flood loadings, is proposed and analysed. These solutions are achieved by using the hand lay-up forming technique utilizing glass, carbon or aramid fibers fabrics pre-impregnated with thermosetting epoxy, polyester or vynilester resins.

The application of these FRP composites is carried out on reinforced concrete columns and beams as well as on brick masonry works aiming to increase in the overall load bearing capacity, especially against horizontal loads. An improved protection against excessive humidity is also envisaged.

The Finite Elements Method based LUSAS software was used to simulate a partially flooded structure. The numerical modeling was carried out in both the un-strengthened and strengthened conditions of the structure in order to assess the increasing in load and deformation capacities of the structural elements. Volumetric finite elements were used for modeling the concrete and masonry members.

## INTRODUCTION

This investigation deals with the retrofitting methods applied to buildings and structures in flood hazard areas especially dry flood proofing. Retrofitting solutions intend to eliminate and reduce the possibility of flood damage. The retrofitting actions are intervention measures such as: elevation of building, relocation of structure outside of floodplain, dry flood proofing through strengthening of the structural elements, wet flood proofing, floodwalls or levees protection.

An example of a flooded house exposed to hydrostatic pressure increasing with the water depth as well as the buoyancy forces is illustrated in Figure. 1. RC

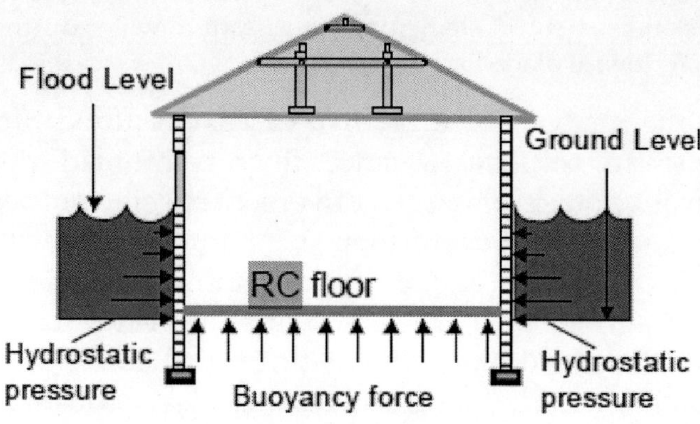


**Figure. 1.** – Hydrostatic pressure acting on masonry walls and reinforced concrete floor.

A suitable protection method will be analyzed in the case of structures placed in areas susceptible flooding. In fig. 2 a series of practical solutions for protection against flooding is shown.

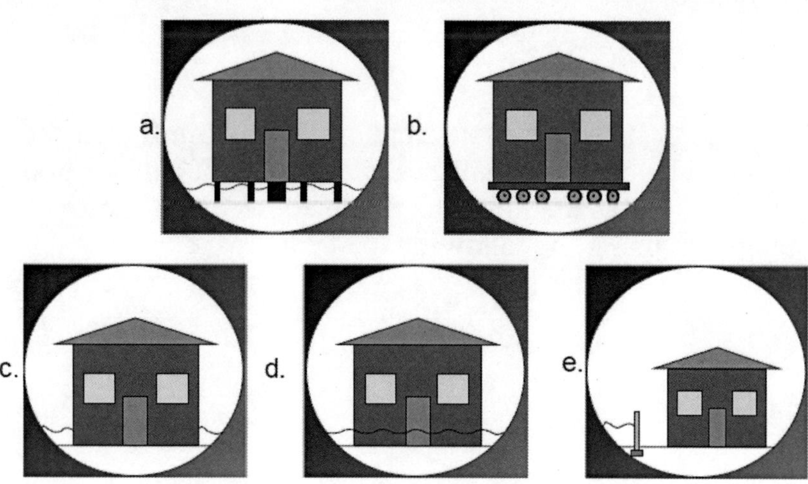


**Figure. 2.** – Methods of protection applied to houses in the case of flood disaster: a – elevation of an existing house using solid walls or columns; b – relocation of an existing building outside of flooded area; c – dry flood proofing

using strengthening methods of structural elements; d - wet proofing where the utilities and structural elements are resistant to water during flooding; e - floodwalls or levees placed around the house [1].

From the analysis of these five cases, dry flood proofing using FRP has been considered since fiber reinforced polymer have particular advantages namely: convenient specific properties (impact absorption, damping performance, fatigue resistance), long term durability (corrosion resistance, environmental resistance), design flexibility (tailored properties, good architectural features), weight saving (dead load reduction), insulating properties (electrical, thermal and acoustic), acceptable fire resistance (low smoke and toxicity). Thus, FRPs can be used in a variety of structural applications such as: repair and strengthening of RC structures, masonry, wooden elements or connections, steel or cast iron elements, etc.

Some possible application solutions for strengthening of structures or structural elements with FRP composites are the following:

a) Visible from outside intervention using plate bonding, hand lay-up techniques, confining with FRP fabrics, etc;
b) Hidden using glass FRP glued sheets or rods inserted in preformed slots or FRP profiles or bars.

## PROPOSED MODEL

An in filled masonry construction with reinforced concrete columns and beams was analyzed in the event of a major flood. The hydrostatical water loading was modeled according to the ASCE/SEI 7 code [2], using the following load combinations:

$$
\begin{aligned}
&1.\ 1.4(D+F);\\
&2.\ 1.2(D+F+T)+1.6(L+H)+0.5(L_r \text{ or } S \text{ or } R);\\
&3.\ 1.2D+1.6W+2.0F_a+L+0.5(L_r \text{ or } S \text{ or } R);\\
&4.\ 0.9D+1.6W+2.0F_a+1.6H,
\end{aligned}
\tag{1}
$$

where: D is the dead load, F – the load due to fluids with well-defined pressures and maximum depths, T – the self-straining force, L – the live load, H – the load due to lateral soil pressure or ground water pressure, Lr – the roof live load, S – the snow load, R – the rain load, W - the wind load and Fa – the flood load.

# THE ANALYZED MODEL

The first step in the performed analysis was to identify the most severe loaded panel of the considered structure. A typical Romanian masonry structure was modeled together with the reinforced concrete framing elements specified by recent design codes (tie-beams and columns - Fig.3).

After this step, the frontal panel was considered independently further on in order to decrease the number of finite elements and reduce the computation time to a reasonable period. Fig. 5 and 6 illustrate the bending moment distribution on the tension face of the panel together with the displacements in the acting direction of the pressing water

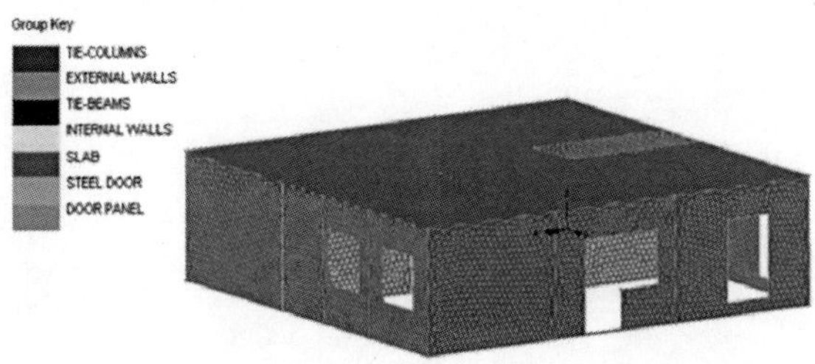


**Figure.** 3. - Finite Elements Mesh in the structural model.

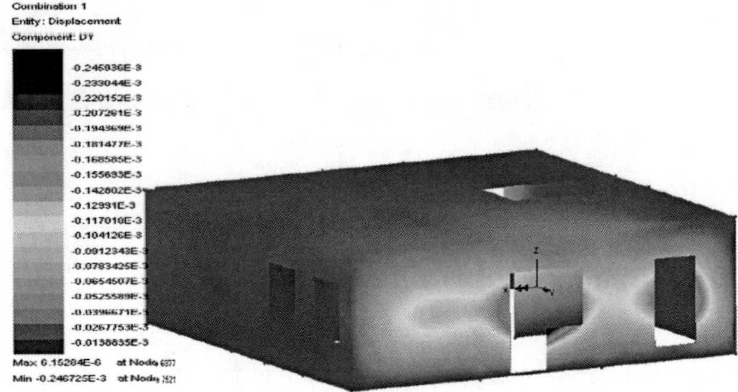


**Figure.** 4. - Displacements map on the model (water action considered in the load combination)

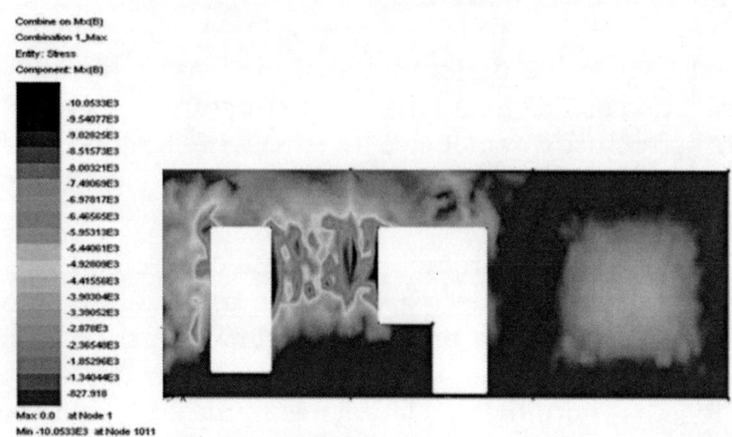


**Figure. 5.** – Bending moment distribution on the interior face of the

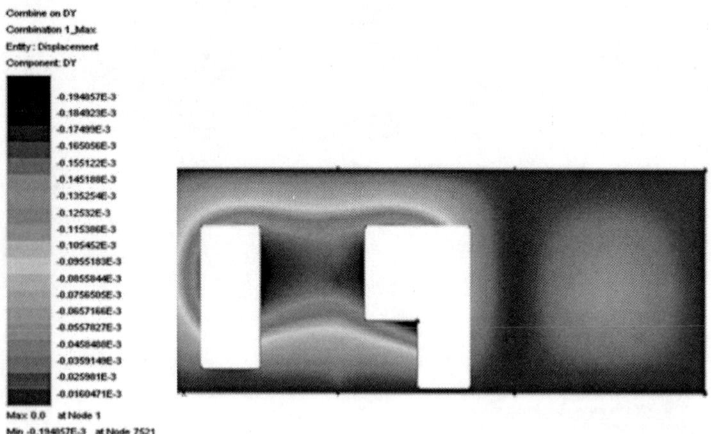


**Figure. 6.** – Displacements in the direction of water pressure action individual panel

The strengthening effect exerted by the fiber reinforced composite material was modeled considering a continuous membrane attached on the interior face of the panel.

A substantial increase in the deformation capacity of the panel was observed. Figs. 7 and 8 show the behavioral differences between the two considered situations (unreinforced and reinforced masonry panels).

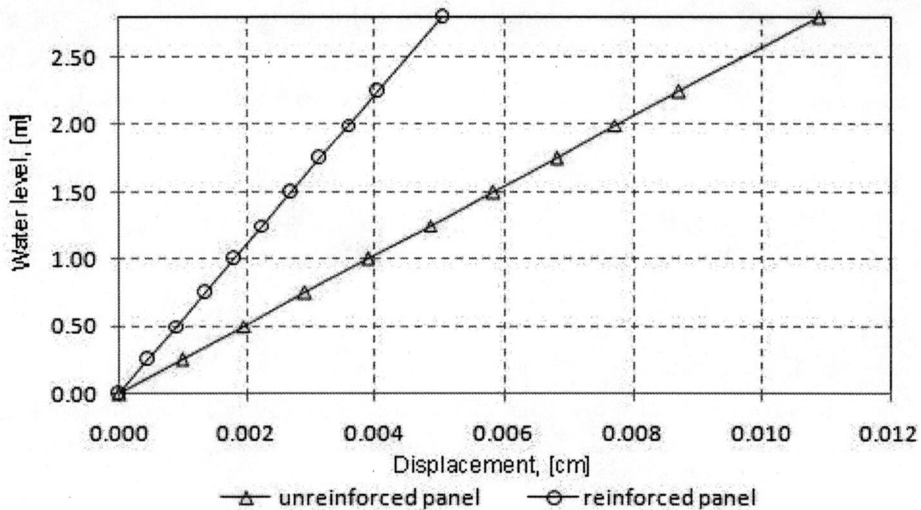


Figure. 7. - Water level *vs*. displacement (centre point of the panel).

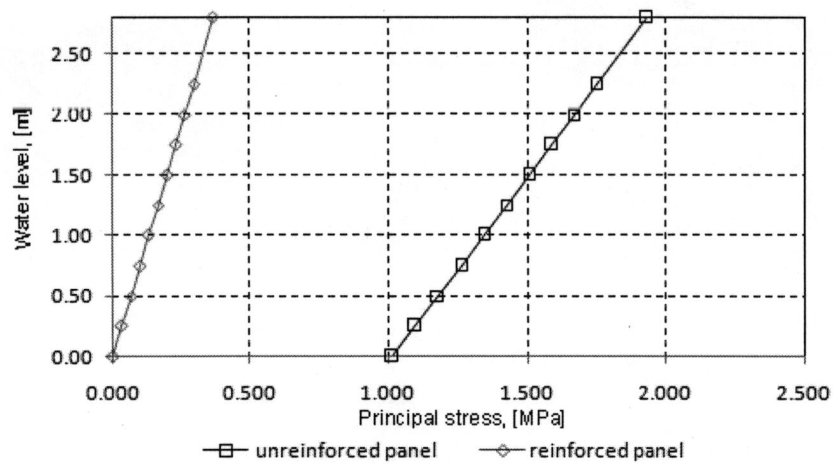


Figure. 8. - Water level *vs*. principal stress (centre point of the panel)

The following conclusions can be formulated after performing the finite elements analysis:

a) Fibre reinforced polymer composites can be successfully utilized to improve the load-carrying capacity of the construction members and structures made of traditional materials, when they are subjected to loading from flooding.
b) Finite element analysis and modeling are suitable tools for analysis of masonry structures strengthened with FRP composite membranes.
c) An important increase in the flexural capacity and stiffness improvement of reinforced masonry panel is possible when FRP composites strengthening systems are used;

## REFERENCES

1. *Engineering Principles and Practices for Retrofitting Flood-Prone Residential Structures,* F.E.M.A. (Federal Emergency Management Agency) January 1995.
2. *Minimum Design Loads for Buildings and other Structures,* American Society of Civil Engineers, ASCE 7-05, 2006.
3. *LUSAS Modeller User Manual" for Civil, Structural and Bridge Engineering,* Version 13.6, 2004-2005, Surrey, UK.

# Chapter 7

# MODELING FIBER COMPOSITES DURING THE CURE PROCESS FOR PIEZOELECTRIC ACTUATION

Darryl V. Murray, Oliver J. Myers

Department of Mechanical Engineering, Mississippi State University, Starkville, USA

## ABSTRACT

Analytical, numerical, and experimental modeling methods are presented to predict deformation after the cure process of thin unsymmetric laminates for piezoelectric actuation. During fabrication, laminates deform to several post-cure room temperature shapes. Thin cross-ply laminates deform to a circular cylindrical post-cure shape while thicker laminates deform to a saddle shape. Post-cure shapes are dependent on ply orientation, thickness, and material properties. Because, CLT alone does not always predict the correct post-cure room temperature shape of the thin composite laminates, an extension of CLT with the Rayleigh-Ritz technique and potential energies are used to better predict these shapes. Finite element models are used to predict the post-cure room temperature shapes. Thin composite laminates are modeled coupling heat transfer and structural mechanics, which are necessary for modeling the cure process. Modeling the fabrication process captured important data such as residual stresses from the cure process, room temperature shapes, and bi-stability of the composite laminates. To validate these analytical and numerical results, experiments were conducted using macro-fiber composite (MFC) patches for morphing the laminates. The experimental piezoelectric morphing results relate well to analytical and numerical results.

## INTRODUCTION

Smart material systems are vastly becoming an integral part in engineering applications. One of the phenomena used in smart material systems is piezoelectricity. Using unsymmetric bi-stable composites, piezoelectric effects can be implemented to achieve a snap through to the other stable shape of the composite [1]. Unsymmetric laminates are defined as laminates that are configured such that the geometric mid-plane is not a mirror image of the ply configurations above and below the mid-plane. This shape change caused by piezoelectric effects can be coupled with other domains and used as a sensor or actuator. To achieve bi-stability, a fabricated unsymmetric laminate is cured to a certain temperature then cooled to the operating or room temperature [1]. The curing process causes the laminate to deform due to the thermal strain gradient between the layers of the laminate. As an essential part of the piezoelectric system, the cure process and the resulting cured shapes are heavily investigated in the current paper.

After curing, a once flat laminate will deform into one of multiple shapes based on the laminate layup and the suggested cure temperature. Research and analysis are being done to predict the deformations and deflections of laminates. Dano and Hyer (2001) predicted the forces and moments that cause cure deformation using common theories based on displacements and strain fields [2]. Modeling three different lay-ups, the Rayleigh-Ritz method and the principal of work was used to determine these forces. Formulating equations for the first derivation of the strain energy, the reference or mid-plane strains and curvatures were approximated. These parameters were used to determine the in-plane and out-of-plane displacements using strain-displacement relationships. A system of equations was then used to approximate the work due to the thermal strains. The research continued on to applying moments to the laminate after the cure process to cause the laminate to snap through to the other stable shape. This research was done as an extension of Dano and Hyer (1996) previous work based on the same theories [3]. C. R. Bowen (2007) experimented with a cantilever beam of a [0/0/90/90] carbon fiber epoxy and an unsupported laminate of [0/90] for morphing testing and analysis [4]. The specimens were cured to a particular maximum temperature with constant pressure. Both specimens formed to one of the cylindrical shapes after the cure process. Piezoelectric actuation was then implemented using macro-fiber composite patches for electrical and mechanical coupling. Although the specific formulation of the cured deflections were not presented, the results were useful.

Hyer and Jilani (1998) presented a method for predicting the deformation of rectangular laminates cooled from the cure temperature to that of the room temperature [5]. The piezoceramic material named THUNDER was used for investigation. Due to the unsymmetric nature of the THUNDER system, deformation was observed as the system cooled from the cure temperature. After analyses were done, it was stated that the occurrence of one of the two stable shapes depends on the geometry of the laminate. This research consisted of also studying the issue of stability and what constitutes a laminate as being in one of the stable states.

The Classical lamination theory suggests that a laminate cooled from the cure temperature will achieve a saddle shape, the shape between the two stable shapes. This suggestion is not always correct. When cooled from the cure temperature, laminates can also attain one of the stable cylindrical shapes (concave up or down). M. Schlecht, K. Schulte, and Hyer (1995) extended the Classical lamination theory to approximate these shapes then compared these results to finite element analysis [6]. The results were used as parameters for the stress and strains during the snap through action to the other stable shape. It would seem that since geometry plays a role in the cured shape of the laminate, the ply orientation would also be a factor. D.

N. Bettes, A. T. Salo, C. R. Bowen, and H. A. Kim (2010) investigates this along with other parameters effecting the cured shape [7].

Hyer also discusses the subject matter of stress analysis of fiber composites [8]. Hyer postulates many of the techniques used to determine stresses, strains, and displacements for a fiber composite including but not limited to these variables in reference to the cured shape. Numerous valuable techniques and assumptions can be found and have been used for this current research. Jones [9] also discusses the mechanics of composites used in this research and includes Hyer's work on post cure shape change deformation. The importance of the approximation of the cure shape and the cure deformation is the cause of the abundance of research done within this area of unsymmetric laminates.

M. Gigliotti and M. R. Wisnom studied the curvature due to the cure process of an AS4/8552 [0/90] laminate [10]. The cure process was broken down in to three stages and further investigation was done. It was hypothesized that inconsistent thermal properties between plies, chemical shrinkage of resin/epoxy, interaction of the tool or mold, and a non-uniform cure temperature are the four thermal stress-strains mechanisms that cause stress in the laminate during the cure process.

The snap through phenomenon for piezoelectric systems are of interest and will be investigated in further research; the present paper focuses on the curing process and approximating the stresses, strains, and curvatures of the unsymmetric laminate during this process. As Schlecht and Schulte studied the thermal effects of the cure process on room-temperature laminates, this paper establishes analytical and numerical modeling for stress, strain, and curvature prediction after the cure process [6]. Using carbon pre-preg sheets, the plies are cut to the specific ply orientation. Then using an aluminum mold and a releasing agent, the laminate is placed in the mold inside a convection oven. During the cure process, there are four mechanisms that cause curvatures and deformation in the laminate [11]. The development of stress in the cured laminate has a significant effect on the piezoelectric actuation and analysis. There can be more advances on the prediction of these variables for better understanding for the application of smart material systems. To achieve optimum snap through with piezoelectric materials, whether sensing or actuators types, it is best that the laminate possess curvatures and strains necessary for particular applications [12-14].

M. W. Hyer also used strain energy theories and virtual work formulation to predict the post cure displacements and curvatures of the laminate [3,5]. These methods were evaluated using various types of lay-ups and cure temperatures. M. W. Hyer formulated the necessary equations to predict the variables then compared these values to experimental values of the same laminate. The most common lay-up was that of [0/90] T and it was used due to the simplicity of the governing equations. Libo Ren and Azar Parvizi-Majidi investigated the CLT and how the predictions of the post cure shape were not always correct [15]. Studying cross-ply shells, Ren and Parvizi modeled the cure shape using the Rayleigh-Ritz energy method. This method allowed consideration of geometric nonlinearity, a problem with the general CLT. M. Gigliotti and M. R. Wisnom along with K. Potter researched bifurcation of the saddle shape of a [0/90] composite [16]. During this study, the researchers observed the saddle shape was unstable compared to the other two cylindrical shapes. Using a numerical models, they compared the Rayleigh-Ritz method model to the numerical models.

As seen in the literature, bi-stability of the composite laminate is achieved during fabrication. After the cure process, multiple deformation shapes can be observed based on the ply orientation and material [9]. This deformation is due to the thermal strain gradient between the layers of the laminate and cure shrinkage during the cure process [10]. As an essential part of the smart material morphing system, the cure process and the resulting post-cure deformation shapes are heavily investigated. Thin unsymmetric composite laminates will deform to one of two circular cylindrical shapes, with the shape in between these two cylindrical shapes being unstable and called the saddle shape. Work has been done to show that the CLT can only predict that the post-cure room temperature shape of all thin composite laminates will be the saddle shape [17]. A Rayleigh-Ritz technique has been established and used here to study the post-cure deformation shapes. This paper presents the CLT first then builds upon it with the methods of strain energy and virtual work for approximations used by M. W. Hyer, Jones, and others to get better post-cure room temperature shape predictions. During the cure process, there are certain mechanisms which cause the displacement and shape change of the composite [18]. With these assumptions, the CLT is prone to error when determining the post-cure room temperature shape. The numerical models have the capabilities to couple thermal, electrical, and mechanical domains and allows user defined equation integration. This phase of research entails modeling the cure process, which is used to more accurately characterize the piezoelectric attachment and actuation in the numerical models. To validate the techniques and methods used, the analytical and numerical models are compared to experimental results.

## ANALYTICALLY MODELING THE CURE PROCESS

The fabrication and cure process of unsymmetric laminates are essential elements of the smart material morphing structure. Models were developed to achieve a better understanding of these processes for piezoelectric actuation. An analytical model is presented first to predict the post-cure shape and deformation. The analytical model was based on both Hyer's and Vizzini's composite material work [3,19,20]. Numerical analysis was done to compare to the analytical results, and is presented following the analytical models. Multi-physics Finite Element Models were created for numerical modeling of both the fabrication process and piezoelectric actuation [21]. Experimental setup and procedures for fabrication and the cure process are then explained. Analytical, numerical, and experimental results will be shown. A post-cure shape deformation comparison between the models will be shown in later sections.

### Classical Lamination Theory

AS4 carbon fiber pre-impregnated laminae are used for the composite laminates of this study. Due to the thickness of the carbon fiber pre-pregs, the laminates in this study are classified as thin composite laminates where the thickness to length ratio is less than $\frac{1}{20}$. This ratio is important for post-cure deformation shapes, thin laminates will deform to a circular cylindrical shape whereas a thicker laminate will deform to that of the saddle shape. The first model was done with the CLT for composite laminates. For CLT analysis, the following key assumptions of the material are made:

- assume smeared properties through thickness;
- assume perfectly bonded laminae with no defects;
- assume zero bond line thickness between plies (no epoxy between plies);
- assume Kirchhoffs hypothesis is valid; plane sections remain planar and perpendicular sections remain perpendicular.

The first assumption states that the individual fibers of the laminae are homogeneous, not varying with direction. The second and third assumptions are made to eliminate any complications with bonding the laminate. The models of the laminate are assumed to be perfectly bonded with no epoxy layer between the plies. The fourth assumption is made in relation to Kirchhoffs hypothesis. This assumptions states that the strain is linear through the thickness of the laminate. The laminate is then treated as an orthotropic material with three planes x, y, and z (also known as 1, 2, and 3). Treating the laminate as an orthotropic material, nine engineering constants of the material are required including Youngs modulus, Poisson ratio, and shear modulus of the material one in each directional plane. These constants are then used to determine the reduced material stiffnesses of the laminate [20]. In order to observe how material stiffness relates to each individual lamina with a specified orientation, the stressstrain relations need to be transformed. Using a method on invariants, the lamina stress-strain relations can be determined for an arbitrary orientation. In terms of the reduced stiffnesses, these invariants can be calculated

$$I_1 = \frac{Q_{11} + Q_{22} + 2Q_{12}}{4} \tag{1}$$

$$I_2 = \frac{Q_{11} + Q_{22} - 2Q_{12} + 4Q_{66}}{8} \tag{2}$$

$$R_1 = \frac{Q_{11} - Q_{12}}{2} \tag{3}$$

$$R_2 = \frac{Q_{11} + Q_{22} - 2Q_{12} - 4Q_{66}}{8} \tag{4}$$

where $Q_{ij}$ are the reduced material stiffnesses [9]. Using these invariants, the transformed reduced stiffness tensor can now be found:

$$E_{11}(\theta) = I_1 + I_2 + R_1 \cos(2\theta) + R_2 \cos(4\theta) \tag{5}$$

$$E_{22}(\theta) = I_1 + I_2 - R_1 \cos(2\theta) + R_2 \cos(4\theta) \tag{6}$$

$$E_{12}(\theta) = I_1 - I_2 - R_2 \cos(4\theta) \tag{7}$$

$$E_{16}(\theta) = \frac{1}{2} R_1 \sin(2\theta) - R_2 \sin(4\theta) \tag{8}$$

$$E_{66}(\theta) = I_2 - R_2 \cos(4\theta) \tag{9}$$

$$E_{26}(\theta) = \frac{1}{2} R_1 \sin(2\theta) + R_2 \sin(4\theta) \tag{10}$$

where θ is the arbitrary ply orientation. The next step in the analysis is to calculate the forces and moments that occur during the cure process. In order to determine these resultant forces and moments the extensional, bending-extension coupling, and the bending stiffness tensors also named the ABD matrix, must be calculated. Using the rotated stiffness tensor equations and the middle surface lamina thickness, the ABD matrix can be determined for each direction of the laminate [9]. The B tensor component of the ABD matrix, the bending-extension coupling stiffness, implies that if the composite is pulled a bending and/or twisting reaction of the composite must occur also. For the thin cross-ply unsymmetric composites, this bend/twist coupling is not expected to occur. Treating the laminate as an orthotropic material comes from the the $D_{ij}$ component of the ABD matrix which is the bending stiffness of the laminate. If $D_{16}$ $=D_{26} = 0$, then this assumption can be made. Unsymmetric laminates contain this type of coupling stiffness while symmetric laminates do not. The force and moment constitutive equations are given by

$$\begin{Bmatrix} N_{11} \\ N_{22} \\ N_{12} \\ M_{11} \\ M_{22} \\ M_{12} \end{Bmatrix} = \begin{bmatrix} A_{11} & A_{12} & A_{16} & B_{11} & B_{12} & B_{16} \\ A_{12} & A_{22} & A_{26} & B_{12} & B_{22} & B_{26} \\ A_{16} & A_{26} & A_{66} & B_{16} & B_{26} & B_{66} \\ B_{11} & B_{12} & B_{16} & D_{11} & D_{12} & D_{16} \\ B_{12} & B_{22} & B_{26} & D_{12} & D_{22} & D_{26} \\ B_{16} & B_{26} & B_{66} & D_{16} & D_{26} & D_{66} \end{bmatrix} \begin{Bmatrix} \varepsilon_{11}^{o} \\ \varepsilon_{22}^{o} \\ \gamma_{12}^{o} \\ \kappa_{11}^{o} \\ \kappa_{22}^{o} \\ \kappa_{12}^{o} \end{Bmatrix} \quad (11)$$

where N and M are the forces and moments. To calculate these strains and curvatures values, the ABD matrix can be inverted to solve for the desired variables. If the forces and moments were to be determined and the strains and curvatures were known, the simple calculation can be done with the matrices. Continuing the CLT, the laminate is treated as orthotropic material to equate the thermal strains to the coefficients of thermal expansion multiplied by the thermal gradient. The residual stresses and strains arising from the cure process can then be found [20].

## RAYLEIGH-RITZ TECHNIQUE

The CLT predicts a saddle shape will occur each time a thin unsymmetric laminate cools from the elevated cure temperature to room temperature. This is due to the failure to capture reference strain due to the large out-ofplane deformations as the laminate is cooled. The Rayleigh-Ritz technique accounts for the large out-of-plane deformations arising from the room temperature cooling. The maximum states of the potential energy are the unstable equilibrium conditions of the laminate. The minimum states of the potential energy are the conditions needed to characterize the deformations of the laminate cooling to room temperature. Using the same engineering constants, mentioned earlier, the reduced stiffnesses of the laminate are found. The coefficients of thermal expansion are also transformed to the ply orientation not aligned in the principal axis. The process is continued to using the method of invariants mentioned in the earlier section to determine the reduced stiffnesses of the laminate. The thermal strains and stresses are then evaluated using the following equations:

$$\sigma_x^T = Q_{11}\varepsilon_x^T + Q_{12}\varepsilon_y^T + Q_{16}\gamma_{xy}^T \tag{12}$$

$$\sigma_y^T = Q_{12}\varepsilon_x^T + Q_{22}\varepsilon_y^T + Q_{26}\gamma_{xy}^T \tag{13}$$

$$\sigma_{xy}^T = Q_{16}\varepsilon_x^T + Q_{26}\varepsilon_x^T + Q_{66}\varepsilon_x^T \tag{14}$$

where the thermal strains, $\varepsilon^T$, are given by:

$$\varepsilon_x^T = \alpha_x \Delta T \tag{15}$$

$$\varepsilon_y^T = \alpha_y \Delta T \tag{16}$$

$$\gamma_{xy}^T = \alpha_z \Delta T \tag{17}$$

where $\alpha i$ are the coefficients of thermal expansion. The middle surface strains and curvatures can then be approximated. Where u, v, and w are the displacement fields:

$$u^o = cx - \frac{a^2x^3}{6} - \frac{abxy^2}{4} \tag{18}$$

$$v^o = dy - \frac{a^2y^3}{6} - \frac{abx^2y}{4} \tag{19}$$

$$w^o = \frac{1}{2}\left(ax^2 + by^2\right) \tag{20}$$

The stresses and strains are then approximated using

$$\varepsilon_x = \varepsilon_x^o + z\kappa_x^o \tag{21}$$

$$\varepsilon_y = \varepsilon_y^o + z\kappa_y^o \tag{22}$$

$$\gamma_{xy} = \gamma_{xy}^o + z\kappa_{xy}^o \tag{23}$$

$$\sigma_x = Q_{11}\varepsilon_x + Q_{12}\varepsilon_x + Q_{16}\gamma_{xy} - \sigma_x^T \tag{24}$$

$$\sigma_y = Q_{12}\varepsilon_x + Q_{22}\varepsilon_y + Q_{26}\gamma_{xy} + \sigma_y^T \tag{25}$$

$$\sigma_{xy} = Q_{16}\varepsilon_x + Q_{26}\varepsilon_y + Q_{66}\gamma_{xy} - \sigma_{xy}^T \tag{26}$$

The post-cure room temperature shape can then be modeled. This step does not include bonding the MFC actuator, therefore only thermal effects are accounted for. The strain energy or total potential energy is given by:

$$\Pi_1 = \frac{1}{2}\int_{-\frac{Lx}{2}}^{\frac{Lx}{2}}\int_{-\frac{Lx}{2}}^{\frac{Lx}{2}}\int_{z_2}^{z_o}\Big[\left(\sigma_x - \sigma_x^T\right)\varepsilon_x + \left(\sigma_y - \sigma_y^T\right)\varepsilon_y + \left(\sigma_{xy} - \sigma_{xy}^T\right)\gamma_{xy}\Big]\mathrm{d}x\mathrm{d}y\mathrm{d}z \tag{27}$$

Carrying out the integrations in equation 27, the potential energy is then reduced to an algebraic equation in terms of the coefficients a, b, c, and d. Once the integrations are computed, Equation (27) will show that the potential energy is also in terms of material properties and geometries of the laminate. These coefficients are then determined by solving the nonlinear algebraic equations that is reduced from equating to zero the first variation of the potential energy with respect to the coefficients,

$$\frac{\partial \Pi}{\partial C_i} = 0, i = 1,\cdots,4 \tag{28}$$

where C is a, b, c, and d. The undetermined coefficients were found using Newton's iterative method. These solutions relate to the equilibrium shapes of the cooled room temperature laminate. To check stability of the solution for the laminate, the second variation of the potential energy must be positive definite.

The following is an example of the results found for the analytical methods mentioned for the post-cure room temperature shapes of the laminates. Table 1 shows the material properties of the carbon fiber used.

The laminates studied were [0/90], [$0_2$/$90_2$], [0/45], and [45/90] square and rectangular laminates. When cooling to room temperature, the cross-ply laminates will deform to one of three possible equilibrium shapes. Two being stable cylindrical shapes and one being the unstable saddle shape, the shape between the two cylindrical shapes. After analytical formulation, these equilibrium shapes can be predicted and are shown in the meshes of **Figure 1** for the [0/90] laminate.

**Table 1**. Material properties of AS4 carbon fiber.

| AS4/3501 Carbon Fiber | |
|---|---|
| $E_1$ | 150 GPa |
| $E_2$ | 20 GPa |
| $v_{12}$ | 0.25 |
| $a_1$ | $-7.815e^{-10}$ 1/K |
| $a_2$ | $6.252e^{-8}$ 1/K |
| $G_{12}$ | 5 GPa |
| Curie Temperature | 449.817 K |

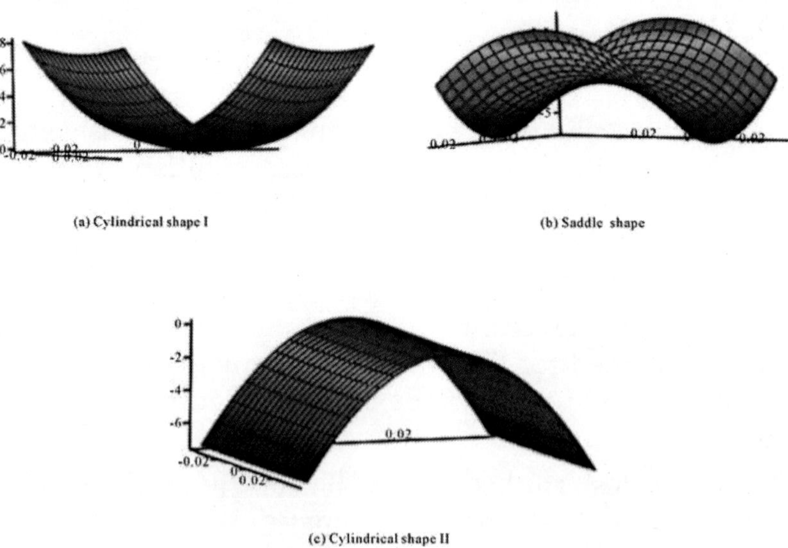


**Figure 1**. [0/90] Cooled room temperature shapes.

As seen in plot two of **Figure 1**, the saddle shape can be referred to as the shape between the two cylindrical shapes. For the simple cross-ply laminates, this shape should not occur after cooling to room temperature. The different orthogonal cylindrical curvatures can be seen and compared in plots one and three. For the [0/90] laminate, the expected post-cure deformation shape is the shape in plot one. After piezoelectric actuation, the [0/90] laminate should snap through to the second cylindrical shape two, shown in plot three of **Figure 1**. The transverse displacement after the curing process was also studied. This was done to investigate the curvatures of both the post-cure shape and the later piezoelectric actuated shapes. For the first cylindrical shape, a (expected shape after curing), the transverse curvature is along side length Lx. This is the curvature for the ideal situation. The third plot shows the second room temperature cylindrical shape of the [0/90] laminate.

Next, predictions can be made for the $[0_2/90_2]$ laminate using the same techniques as described earlier. With more plies than the [0/90] laminate, the curvature is expected to be different due to the increase of plies. The transverse curvature for the first cylindrical shape (expected shape after curing) of the $[0_2/90_2]$ laminate is expected to be along the length of side $L_y$. When compared with the [0/90] laminate this curvature is in opposite the direction along the other axis. The equilibrium shapes, shown as meshes, of the $[0_2/90_2]$ laminate in **Figure 2** confirms the different post-cure shape of the laminates.

The first plot (a) in **Figure 2** is the expected post-cure shape of the $[0_2/90_2]$ laminate. Plot b represents the shape between the two cylindrical shapes, the saddle shape. Plot c shows the expected actuated displacement shape of the $[0_2/90_2]$ laminate.

The [0/45] laminate was modeled using the RayleighRitz technique as described above. With a ply orientation at 45°, it is expected that the laminate will not deform to one of the equilibrium cylindrical stable shapes. Instead, the [0/45] laminate should deform to the saddle shape. As research shows, this shape is unstable compared to the simple cross-ply laminates. The stability is determined by the second variation of the total potential energy, or the second derivative of the Jacobian matrix. If this matrix is positive definite, which confirms that the equilibrium solution is stable. As shown in **Figure 3**, the analytical models predicted the post-cure shape of the [0/45] laminate will deform to the saddle shape. The meshes in **Figure 3** may seem flat when compared to the simple cross-ply laminates, but that is due to the instability of the laminate.

The curvature of the [0/45] laminate is along both the x and y axis. This is a characteristic of the saddle shape. After fabrication, the [0/45] laminate should deform to this saddle shape. For piezoelectric actuation the [0/45] laminate is still going to snap through to another saddle shape (shown in later chapters), but unlike the simple

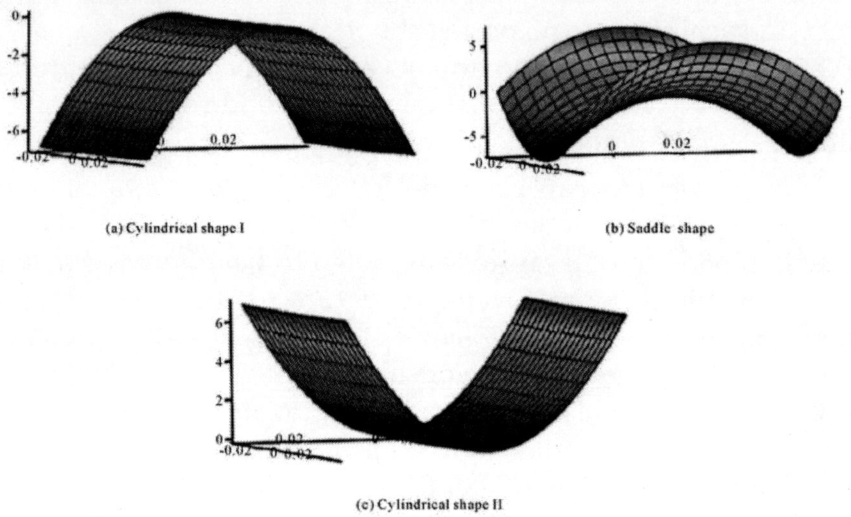


**Figure 2**. $[0_2/90_2]$ Cooled room temperature shapes.

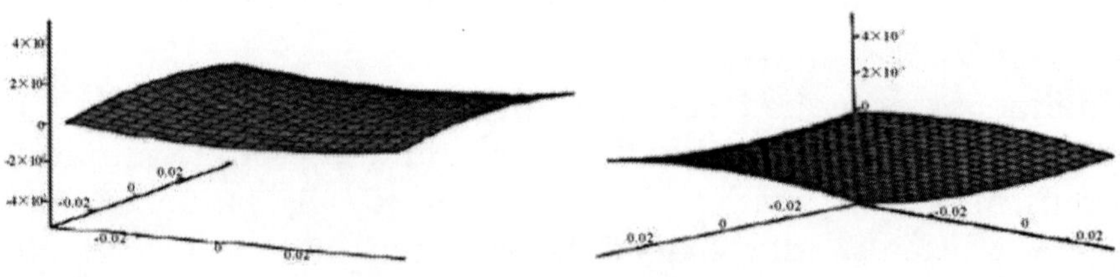


**Figure 3**. [0/45] Cooled room temperature shapes.

cross-ply laminates, due to the instability of the [0/45] laminate removal of the snap through force will cause the laminate to snap back to its original shape shown in Figure 3. These analytical results can now be used as predictions of the post-cure room temperature shapes as well as models to help accurately capture the effects of the piezoelectric actuation models. Comparisons were made to the numerical models and the experimental results.

## NUMERICAL MODELING THE CURE PROCESS

Multi-physics numerical FEA models were used to compare analytical models and experimental results. The present models required implementing the Thermal-Structural Interaction mode, which couples both, the structural mechanics and the heat transfer modules. The analysis type of the stress-strain and the heat transfer modules were chosen as static and transient types respectively. Due to the small thicknesses of the laminates studied, models were done using linear and quadratic mesh element types.

### Initial Setup

The Thermal-Structural interaction mode allowed for including thermal expansion due to the coefficients of thermal expansion(CTE) during the cure process. The CTE's along with geometry and other material parameters, play key roles during the heating and cooling of the laminate. In numerical model 3-D space, new coordinate systems are needed to specify certain fiber orientations for plies not oriented in the global or principal material axes. This feature is done by rotating the global coordinate system from the original x, y, z axis to the angle needed for the specific ply. Allowing $0^o$ to represent the default or global coordinate system (no need to change coordinate system if ply is oriented at $0^o$), the remaining plies coordinate systems can be changed using the consecutive rotation axes option. After the laminate is drawn with the specified dimensions, the material properties and conditions can be specified in the structural mechanics subdomain settings of the model. The heat transfer subdomain settings controls the heat source and the thermal properties of the laminate. The structural domain settings control the material properties of the laminate, including the fiber orientation.

The laminate model was treated as an orthotropic material as stated earlier. In these settings, the necessary engineering constants along with the selection of the global coordinate system and specified created coordinate systems can be entered. The initial pressure of the laminate can be selected. An initial pressure was used to model the mold pressing the laminate flat during the heating of the laminate. Similar to the sub-domain settings, there are separate boundary condition settings for the structural application and the heat transfer application. Modeling free thermal expansion, there were no constraints on the boundaries or edges of the laminate. Compared with the analytical model, this assumption can be made due to the absence of external loads during the cooling of the laminate to room temperature. When solving the model, the solver parameters are important for accuracy and solver convergence. For the structural application mode, the solver was chosen based on the static analysis nature of the problem. This analysis type differs from the heat transfer application mode, however. For the heat transfer application mode, the analysis was chosen to be transient based on the steps of the curing process. Therefore, the analysis type needed to couple both the static and transient nature of the curing process.

## Coupling Structural Mechanics and Heat Transfer

The numerical analysis allowed for more complex lamina behavior than the CLT assumptions used in the analytical modeling. For the structural mechanic analysis mode of the numerical modeling, a few important equations were used to define the domain of the laminate. Using the solid, stress-strain application for 3-D modeling, the straindisplacement relationship equations were specified as follows:

$$\varepsilon_x = \frac{\partial u}{dx} \tag{29}$$

$$\varepsilon_y = \frac{\partial v}{\partial y} \tag{30}$$

$$\varepsilon_z = \frac{\partial w}{\partial z} \tag{31}$$

$$\varepsilon_{xy} = \frac{1}{2}\left(\frac{\partial u}{\partial y} + \frac{\partial u}{\partial x}\right) \tag{32}$$

$$\varepsilon_{xy} = \frac{1}{2}\left(\frac{\partial v}{\partial z} + \frac{\partial w}{\partial y}\right) \tag{33}$$

$$\varepsilon_{xz} = \frac{1}{2}\left(\frac{\partial u}{\partial z} + \frac{\partial w}{\partial x}\right) \tag{34}$$

The stress-strain relation of the numerical analysis is given by the generalized Hooke's Law formulation:

$$\sigma = D\varepsilon \tag{35}$$

where D is the stiffness matrix of the material. The Structural Mechanics mode is based on the weak formulation of equilibrium equations in the global stress components. For 3-D models, the equilibrium equation is represented by:

$$-\nabla\sigma = F \tag{36}$$

where $\sigma$ is the stress tensor and F is the volume or body forces [21].

The sub-domain setting for the heat transfer mode was specified using Fourier's law:

$$-\nabla(k\nabla T) = Q \tag{37}$$

where T is the temperature being measured at that instant, Q is the heat source, and k is the thermal conductivity. Specifying the heat transfer boundary equation using the heat flux equation,

$$n(k\nabla T) = q_o + h(T_{\text{inf}} - T) + \text{Const}\left(T_{amb}^4 - T^4\right) \tag{38}$$

where $q_o$ is the inward heat flux of the composite being modeled, h is the heat transfer coefficient, $T_{inf}$ is the external temperature, Const is the numerical constant for heat transfer relations, and $T_{amb}$ is the ambient temperature. The heat transfer boundary equation is a form of the heat convection equation. The second half of the heat transfer boundary equation is the heat transfer due to radiation. Although shown, it was not used in the models. The equations in this section are all derived or are a form of the same analytical equations used in the numerical analysis. Using the same material properties as in **Table 1**, the laminates were modeled and the post-cure room temperature shapes were predicted.

Applying an initial pressure of 5.861 × $10^5$ Pa to the laminate models the press applied from the mold during the heating of the laminate. The laminate is then heated to 449.817 K then cooled to room temperature, 296.483 K. **Figure 4** shows the numerical solutions of the postcure room temperature shape for the [0/90] rectangular and square laminates. It is seen that for both [0/90] laminate geometries, the symmetric circular transverse curvature is in the negative z-direction. This curvature can be seen in the first plot of **Figure 4** along the x-direction and along the y-direction in the second plot.

The [0/90] laminate was modeled with a total thickness of 0.36002 mm with square length of 76.2 mm and rectangular lengths of $L_x$ = 61 mm and $L_y$ = 120 mm which classified the laminates as thin laminates. In addition to being thin laminates, they were simple cross-ply laminates therefore the saddle shape is not present here. Increasing the number of plies to a $[0_2/90_2]$ laminate, different cylindrical room temperature shapes are observed. **Figure 5** is the numerical solution of the room temperature shape of the $[0_2/90_2]$ laminate. The $[0_2/90_2]$ laminate was modeled at 0.72004 mm total thickness. The lengths were the same as the [0/90] laminate. Therefore the saddle shape is not seen here. For the $[0_2/90_2]$ laminate, the symmetric circular transverse curvature is in the positive z-direction. This curvature is seen in the first and second plot of **Figure 5** along the y-direction.

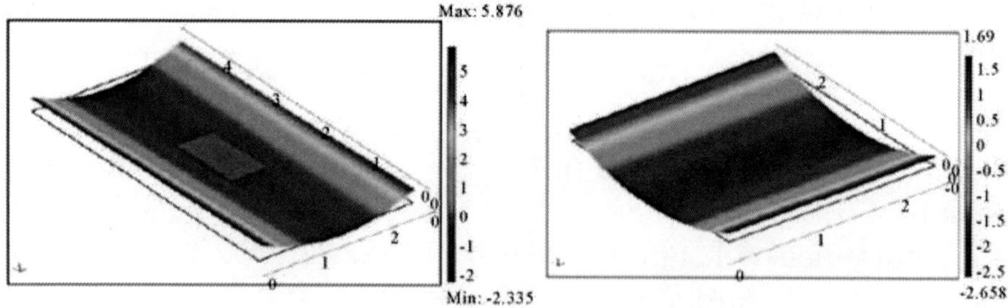


**Figure 4**. Rectangular and square [0/90] cylindrical shape I.

For laminates that are not simple cross-ply laminates, a saddle shape as the room temperature post-cure shape is expected. Experimental and analytical work [6] has been

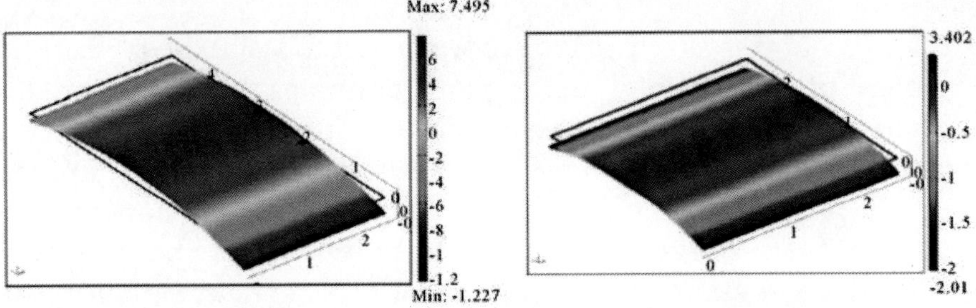


**Figure 5**. Rectangular and square $[0_2/90_2]$ cylindrical shape I.

done to better understand the saddle shape and the instability of the laminate. The simple cross-ply laminates deform while cooling to room temperature to a symmetric circular shape. This is seen in the numerical solutions above. More complex laminates will deform unsymmetrically with a twist curvature similar to the saddle shape. This twist curvature develops as the laminate cools to room temperature. Applying a force to this saddle shape will cause deformation, but unlike the simple cross-ply laminates, removal of the snap through force will cause the laminate to return to its original shape. This classifies the laminate in the saddle shape as unstable. For this study, [0/45] and [45/90] were used for experimentation. Using the same parameters as the simple cross-ply laminates, the [0/45] laminates were modeled.

**Figure 6** shows the rectangular and square [0/45] laminate at room temperature. The saddle shape is observed due to the 45° ply. The twist curvature which characterizes the saddle shape is seen in all three plots. For the rectangular [0/45] laminate in plot 1, two edges deform upward in the positive z-direction while the other two edges deform downward in the negative z-direction.

# EXPERIMENTAL FABRICATION AND RESULTS

## Experimental Fabrication

A standard hand lay-up procedure was used to stack the plies of the laminate. Both square and rectangular laminates were fabricated to better understand the role of the material properties and geometry in piezoelectric actuation. A mold was designed to press the laminate flat while heating to the cure temperature. A convection oven at Mississippi State University Raspet Flight Research Laboratory was used for curing the composite laminates. The mold held the composite laminate flat while in the oven at a cure temperature of 449.817 K. Unsymmetric laminates and the snap through phenomenon work best when the laminate can be classified as a thin laminate.

After removal of the mold from the oven, the press is released allowing the laminate to deform freely. Thin laminates, where the length and width are large compared to the thickness, will deform to one of two cylindrical shapes as discussed earlier [8]. The laminate is then left over night to complete cooling. **Figure 7** shows both the [0/90] and $[0_2/90_2]$ laminate after the cure process. The circular curvature seen in **Figure 7** for the [0/90] laminate and the $[0_2/90_2]$ is similar to the prediction of the analytical and numerical models.

The post-cure transverse curvature in the $[0_2/90_2]$ is greater than the curvature of the [0/90] laminate. This is possible due to the increase of the number of plies in the laminate. Although the transverse curvature is greater, with more plies the laminate becomes stiffer, therefore making it more difficult to apply force for snap through without damaging the laminate.

The next composites fabricated were the laminates that were not classified as cross-ply laminate. These laminates, as discussed earlier, is expected to deform to a saddle shape. The twist curvature that arose from the cooling of the laminate is seen in both the [0/45] and [45/90] laminates in **Figure 8**. The curvature is similar to numerical models presented earlier.

For comparisons seen in later sections, the transverse displacement of the deformed laminates were measured. Due to the symmetric curvature of the simple cross-ply laminates, $[0_n/90_n]$

where n is the number of plies, this symmetric cylindrical shape can be measured using a FARO gage Coordinate-measuring machine (CMM). The CMM establishes a coordinate system for the laminate. This coordinate system is then used to measure the deformation and mesh the post-cure room temperature laminate. The difference between the post-cure room temperature symmetries of the cross-ply laminates and other laminates can now visibly be seen. For the cross-ply laminates, the post-cure room temperature curvatures were symmetric as expected. For the [0/45] and [45/90] laminates, the curvature is not symmetric as the cross-ply laminates. This was expected and predicted due to the

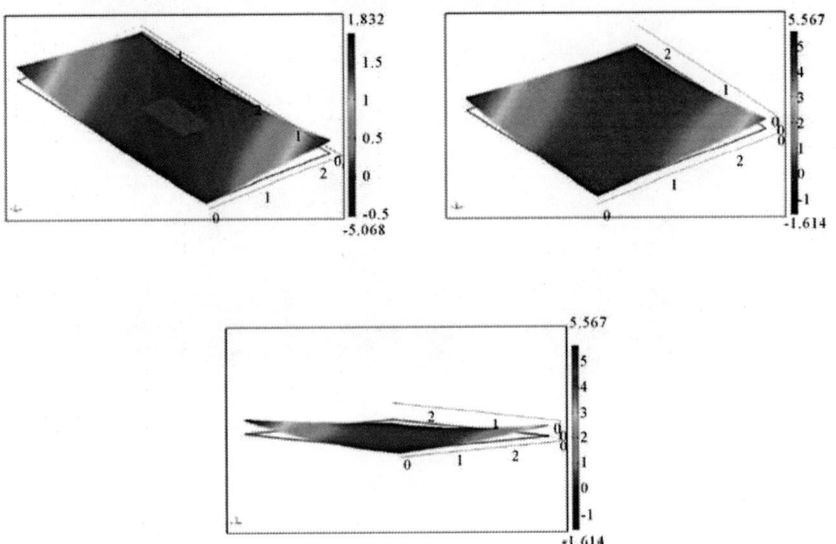


**Figure 6**. Rectangular and square [0/45] shape I.

ply orientation of the laminates. This saddle shape when compared to the cylindrical shapes of the cross-ply laminates is classified as an unstable shape. Applying a force to the post-cure room temperature shapes causes the laminates to snap through to another shape. For the crossply laminates this shape is the second cylindrical shape seen in previous sections. After snap through, this shape remains upon removal of the snap through force, therefore being a stable shape. For the laminates not classified as cross-ply laminates, the laminate will snap back to its original (post-cure room temperature) shape upon removal of the snap through force. Therefore these laminates are unstable.

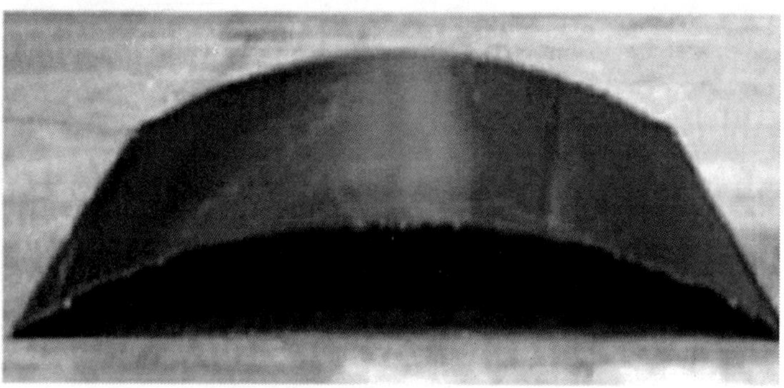

**Figure 7**. Cured room temperature shapes of [0/90] and [$0_2$/$90_2$].

**Figure 8**. Cured room temperature shapes of [0/45] and [45/90].

## EXPERIMENTAL COMPARISONS

For the analytical method, the Rayleigh-Ritz technique along with the total potential energy was used to predict the post-cure shape of the thin laminates. After fabrication, the profile of the laminates were measured using the CMM gage instrument. This instrument provided a coordinate system and data points for the deformed laminate, which provided meshes of the laminates. The data points can also be compared to the transverse post-cure curvatures of the analytical and numerical models. Because laminates wet thin and the CMM required contact, laminates were easily moved vertically by the gage arm of the CMM. When collecting the data points of the post-cure shape, the CMM arm would forcefully displace the laminate up or down. This caused a few errors when plotting the data points of the laminates from the CMM instrument. Laminates were taped to the measurement table to increase stiffness and minimize deflection to the the CMM. Although this provided better control of the arm and laminates, the error was still evident as the CMM arm moved across the laminate.

The post-cure fabrication results of the [0/90] rectangular laminate is shown in **Figure 9** below.

The comparisons for the experimental results and the numerical model are shown in **Figure 9**. The solid black line represents the experimental results and the red traingles represent the numerical model. In plot one, the curvature is given along the x-axis from end to end. The coordinate system here places the origin at the edge of the laminate. Although reasonably close, the error seen is due to the probe of the FARO CMM arm. The same results can be shown for the square [0/90] laminate in Figure 10. As expected, the post-cure curvatures for the [0/90] laminate are in the negative z-direction along the x-axis. This post-cure curvature is expected to be opposite for the $[0_2/90_2]$ laminates.

The $[0_2/90_2]$ laminates have been compared similar to the [0/90] laminates. As seen in other sections, the postcure curvature of the $[0_2/90_2]$ laminates are in the positive z-direction along the y-axis. The different curvature of the $[0_2/90_2]$ and the [0/90] laminate is due to the more dominate 90° plies. While fabricating the laminates, it

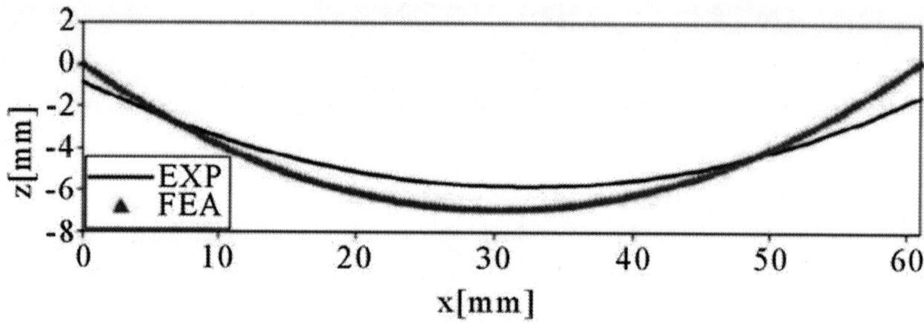


**Figure 9**. Rectangular [0/90] post-cure room temperature displacement.

was observed after cutting to size, that the 90° carbon pre-preg would start to deform to this shape when left alone. Therefore, during cooling of the $[0_2/90_2]$ laminates this deformation is to be expected. Similar to the [0/90] laminate, the numerical model will be shown as red triangles on the plot while the experimental results will be represented by solid black lines in **Figure 11**.

The maximum post-cure curvatures for the simple cross-ply laminates are shown in **Table 2**. The laminates not classified as the simple cross-ply laminates could not be measured with the FARO CMM gage due to the unsymmetric post-cure curvature. Therefore the other laminates were investigated by the use of numerical models and experimentation. The analytical and numerical models can now serve as preliminary models necessary for piezoelectric actuation.

The comparisons of the $[0_2/90_2]$ laminate for the numerical model and the experimental results are shown in plot one of **Figure**

11. The curvature shows that the maximum magnitude for the post-cure transverse displacement is going to occur at the center laminate, approximately by 7.5 mm. The same comparisons are made for the square $[0_2/90_2]$ laminate.

## PIEZOELECTRIC ACTUATION

Using the results of the sequential models of the fabrication process, the effects of piezoelectric actuation were

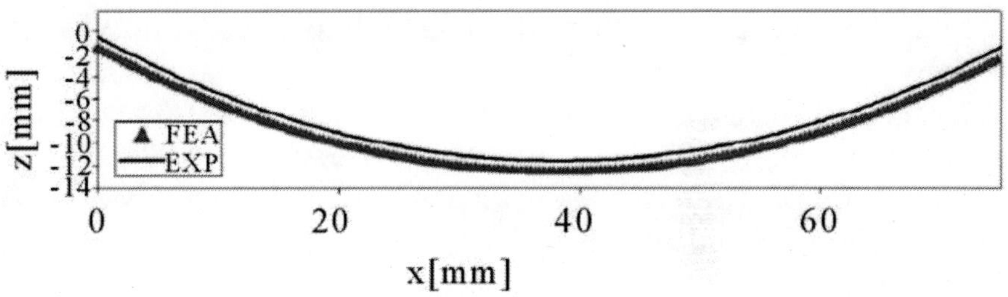


**Figure 10**. Square [0/90] post-cure room temperature displacement.

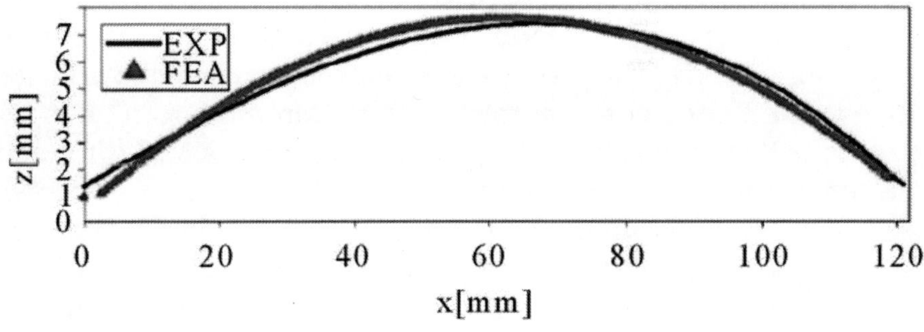


**Figure 11**. Rectangular $[0_2/90_2]$ post-cure room temperature displacement.

**Table 2**. Maximum post-cure displacement curvatures (mm).

| | [0/90] rect. | [0/90] square | $[0_2/90_2]$ rect. | $[0_2/90_2]$ square |
|---|---|---|---|---|
| Analytical | 7.177 | 9.232 | 6.0148 | 7.477 |
| Numerical | 6.342 | 10.632 | 6.02 | 8.21 |
| Experimental | 5.649 | 10.711 | 5.716 | 6.552 |

captured. These preliminary models enhance the knowledge of the key parameters needed for piezoelectric actuation. The MFC actuator used for experimentation is distributed by Smart Material Corporation. The MFC was invented by NASA in 1996 and since then has been continued to be improved for many applications. The piezoelectric material of the MFC actuator are r piezoceramic rods of the actuator [1]. These rods are sandwiched between electrodes and polyimide film as shown in **Figure 14**. The electrodes are in an interdigitated pattern allowing for direct application of the a voltage to the rods, which is the converse piezoelectric effect. Once the voltage is applied, the actuator will elongate or deform. This system enables in-plane poling and actuation, both which are used for this research. The operation mode used for experimentation was the $d_{33}$ operating mode. This mode elongates the rods when a voltage is applied. The tensor notation for the operation mode can be broken down to better understand the proper applications and differences between other operation modes. The first variable, 3, represents the axis of polarization for the material. This direction 3, is through the thickness of the material also known as the z-direction if in x, y, z coordinates. Making the 1 and 2 directions align in the plane of the piezoelectric material. The second variable, 3, represents the direction of which the state of the piezoelectric material should be analyzed [22]. With the $d_{33}$ operation mode and the assumption of free strain, the constitutive equations are reduced. Free strain is the strain produced in the piezoelectric material when there is no resistive stress on the material. Piezoelectric actuation will be shown for the [0/90] laminates.

## Analytical Morphing Models

For piezoelectric analysis, the plates or in this case MFC actuators, were classified as thin plates. Thin plates are those with a ratio of thickness to length less than 1/20. This allows the complex three-dimensional problem to be reduced to a problem in two dimensions [22]. Using the same assumptions of the classical plate theory, the thin plate is now classified as a plane stress problem. The analysis implements the total potential energy approach similar to the post-cure shape analytical work. The Ritz method is used for the variational principal solution and is reduced to a set of linearly independent equations by an assumed form of generalized coordinates. This same procedure was done in previous sections for modeling the cure process. The post-cure room temperature shapes in previous sections had to be accounted for in piezoelectric analysis. When using the total potential energy to model the post-cure shapes, the driving force was the heat transfer taking place. Now for piezoelectric analysis, the driving force will be piezoelectricity. The total potential energy in response to the MFC actuation is given by:

$$\Pi_2 = \frac{1}{2}\int_{-\frac{LxMFC}{2}}^{\frac{LxMFC}{2}}\int_{-\frac{LyMFC}{2}}^{\frac{LyMFC}{2}}\int_{z_2}^{z_3}\Big[\left(\sigma_x^a - \sigma_x^{Es}\right)\varepsilon_x + \left(\sigma_y^a - \sigma_y^{Es}\right)\varepsilon_y + \left(\sigma_{xy}^a - \sigma_{xy}^{Es}\right)\gamma_{xy}\Big]\mathrm{d}x\mathrm{d}y\mathrm{d}z \tag{39}$$

where $\sigma_x^a, \sigma_y^a, \sigma_{xy}^a, \sigma_x^{Es}, \sigma_y^{Es}, \sigma_{xy}^{Es}$ are the stresses and

$\varepsilon_x, \varepsilon_y$ and $\gamma_{xy}$ are the strain strains in the MFC actuator which include the stresses induced by bonding the MFC actuator to the laminate [23]. The total potential energy for the actuation of the laminate, taking the postcure deformation into account is then given as:

$$\Pi_3 = \Pi_2 + \Pi_1 \tag{40}$$

Using the same techniques, the undetermined coefficients a, b, c, and d can be found by minimizing the total potential energy. Thus giving the actuated shapes for the MFC actuator and laminate. For the post-cure [0/90] laminate, the curvature was in the negative z-direction along the x axis. For the cylindrical shape II, the curvature is expected to be in the positive z-direction along the y axis (this would be known as the orthogonal cylindrical shape). The postcure shape (cylindrical shape I) of the $[0_2/90_2]$ laminate has transverse curvature in the positive z-direction along the yaxis. The orthogonal cylindrical shape to the deformed post-cure shape of the $[0_2/90_2]$ laminate would have curvature in the negative z-direction along the x-axis.

As shown in **Figure 12**, the actuated deformation obtained from the total potential energy method predicted the [0/90] laminate to deform in the positive z-direction. The expected curvature for the $[0_2/90_2]$ laminate has also occurred.

## Actuation Modeling

After the curing process is simulated, the same models can be used to model the piezoelectric actuation. Running the structural mechanics and heat transfer modules together, gave the post-cure room temperature shapes of the cross-ply laminates. Using these existing models, the 3D piezoelectric solid module was added. This allowed for the same post-cure curvature (same geometries from cure process models) to be used for modeling piezoelectric actuation. The same element types were used as in the fabrication models. After drawing the rectangular 27.9 × 13.995 mm MFC actuator at the center of the laminate on the first ply, the laminate and actuator was bonded to the laminate by creating a contact pair between the bottom surface of the MFC actuator and the top surface of the laminate. This was done to make the MFC actuator and laminate active in one domain. The actual MFC actuator has piezoelectric fibers along the length of the actuator, which corresponds to the zero direction. Therefore, for the numerical models, the MFC actuator was bonded to the laminate according to the direction of the first ply. Using the same structural mechanics domain settings from the cure process models, the ply orientation, thickness, and material are recalled for the 3D piezoelectric solid module. These models implement applying the voltage to the MFC actuator using the following piezoelectric equations in the stress-strain form.

$$T = c^E S - e^T E \tag{41}$$

$$D = e^S + \varepsilon^S E \tag{42}$$

where the variables T, c, E, S, e, D, $\varepsilon^S$ are the stress, Young's modulus, electric field, strain, piezoelectric constant, electric displacement, and the permittivity of the piezoelectric material. The Piezoelectric Decoupled, anisotropic model was used for the actuation models. Using this modeling, the full six-by-six elasticity matrix was defined and the boundary conditions are established [21].**Figure 13** shows the simulation results of the [0/90] laminate post-cure and post actuation where you can see the change in shape direction after the MFC has been actuated.

## Actuation Experiments

The bonding process done in this research was the process suggested by the Smart Material Corp. For experimental uses, a LabView chassis was used to control the system. For experimentation, the programming voltage was increased from zero by 0.02 V DC per 45 seconds until snap through was observed. An experimental snapthrough laminate is shown in **Figure 14** and the experimental set-up is shown in **Figure 15**.

After snap through of the laminate to the orthogonal cylindrical shape II, the voltage load is removed to assure the cylindrical shape II is stable. If the shape was unstable, removing the load (voltage) would cause the laminate to snap back to cylindrical shape I. In **Figure 16**, the transverse actuated displacement is presented in the second picture.

After running the experimentation process a number of times and experiencing snap through, it was observed for the rectangular [0/90] laminate, an estimated voltage at snap through was 391 V DC. This voltage magnitude was then used in the numerical model to get better results for comparisons. Comparing to the analytical predictions of the post-cure and actuated shape, the experimental results are similar. The same procedure was done with the square

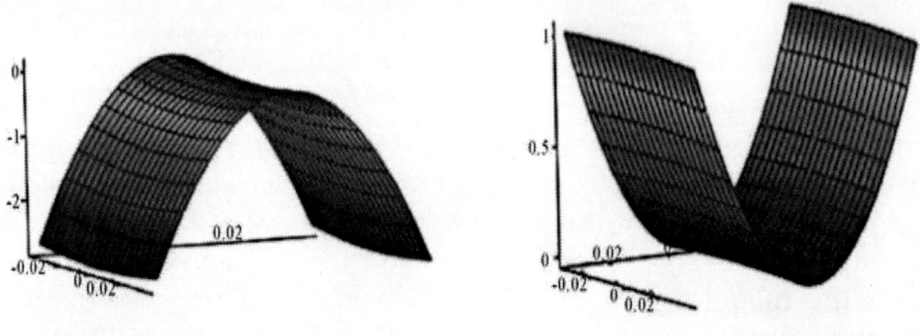


**Figure 12**. Predicted actuated shapes, cylindrical shape II for [0/90] and $[0_2/90_2]$.

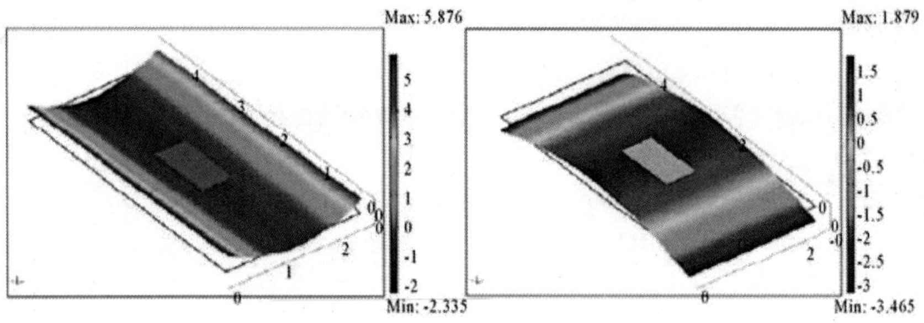


**Figure 13**. Rectangular [0/90] room temperature shape vs actuated shape.

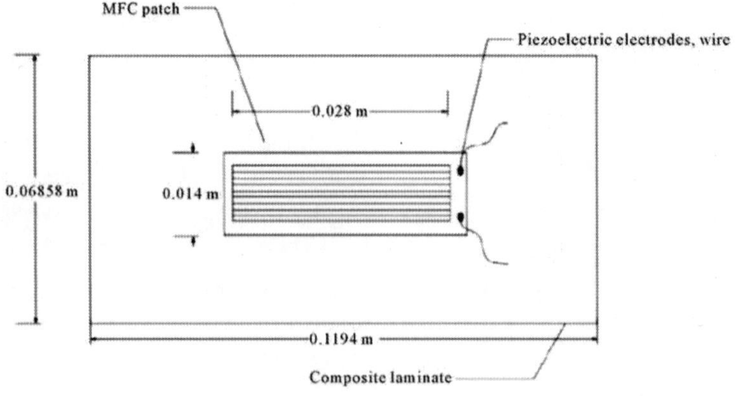


**Figure 14**. Experimental MFC patch on rectangular laminate.

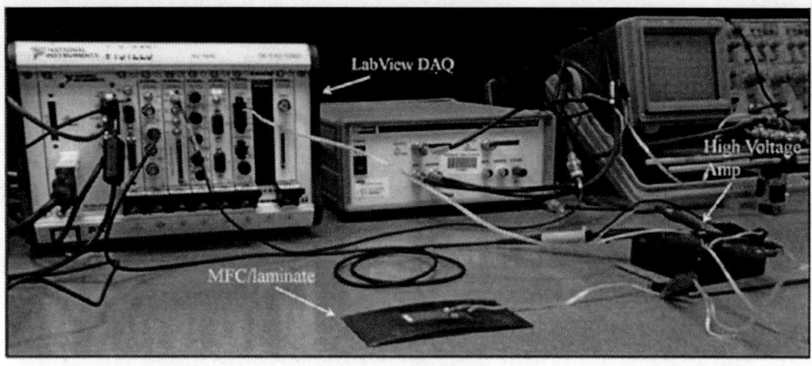


**Figure 15**. Experimental set-up.

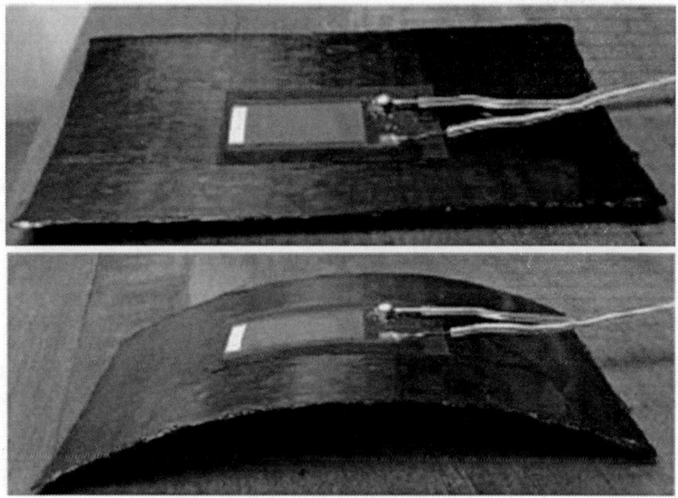

**Figure 16**. Cylindrical shapes I, II of the rectangular [0/90].

[0/90] laminate, seen in **Figure 17**. The estimated voltage at snap through was 951.45 V DC, which was close to three times more than the snap through voltage for the rectangular laminate.

After cooling to room temperature and bonding the MFC actuator to the deformed laminates, the piezoelectric actuation experiments are ready to be done. The [0/90] laminates were done first for piezoelectric actuation. As seen in previous chapters for the [0/90] laminate, the snap through from cylindrical shape I should cause the laminate to have actuated displacement along the y-axis in the positive z-direction, this being cylindrical shape II. The rectangular [0/90] laminate is shown in the **Figure 18**.

As observed in earlier sections, the actuated displacement curvature is predicted by the analytical and numerical models. These predictions are then compared to the experimental results. The rectangular [0/90] laminate comparisons of the numerical and experimental results are reasonably accurate. As expected, the actuated snap through curvature of the rectangular [0/90] laminate is along the y-axis, which is the orthogonal cylindrical shape II. This proves the numerical and analytical models are reasonable prediction models for actuation. The same actuated snap through curvature is expected for the square [0/90] laminate shown in Figures 18 and 19.

Due to the size and flexibility of the laminate, when using the FARO gage CMM to obtain data points the laminate will tend to move. An uncalibrated and unspecified load was required for the FARO gage to capture the deformation date of the flexible laminates. Although a laser profilometer would serve as a better tool for shape and deformation characterization, none were available. The data obtained from the experimental actuation therefore will not be as precise as the FEA ideal models. When comparing the actual experimental result to the FEA model, this error can be seen. Showing similar actuated shapes as the analytical model suggested, the square and rectangular cross-ply laminates behaved as expected. This was done by using the preliminary methods and techniques for modeling the cure process. With the results from those preliminary models, the piezoelectric actuation effect can be more accurately modeled. Modeling the cure process first for preliminary measures of the piezoelectric actuation allowed these methods and techniques to serve as reasonable predictions for the postcure room temperature shapes which is imperative to accurately modeling piezoelectric actuation of unsymmetric laminates.

## CONCLUSION

Models were developed to characterize and predict the post-cure shape of the unsymmetric laminates for piezoelectric actuation. The initial model consisted of the CLT where assumptions of the laminate had to be made for the theory to work. Analysis were done with the CLT

**Figure 17**. Cylindrical shapes I, II of the square [0/90].

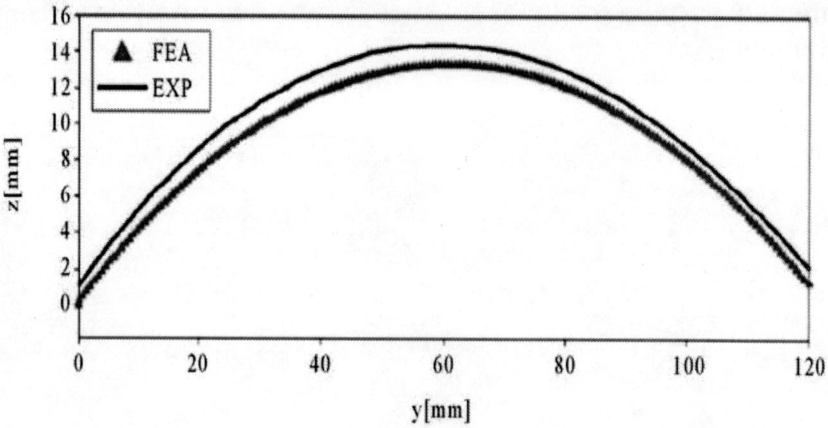


**Figure 18**. Rectangular [0/90] laminate actuated displacement.

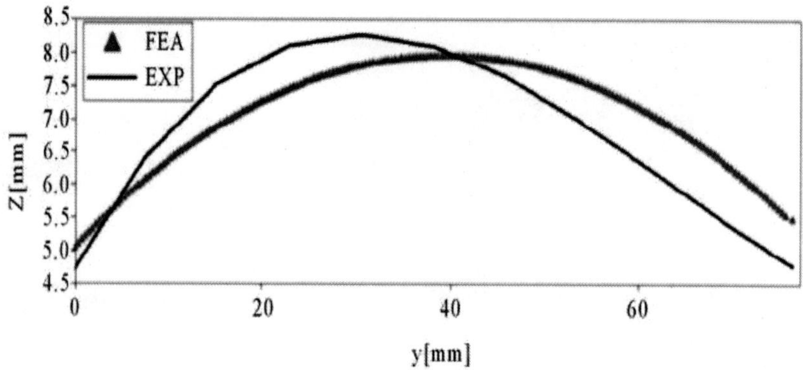


**Figure 19**. Square [0/90] laminate actuated displacement.

alone and it was observed that the laminates were all deforming to the saddle shape when cooling to room temperature. The CLT predicts that the room temperature shape of all unsymmetric laminates will be the saddle shape. Due to the failure to capture reference strains due to the large out-of-plane deformations as the laminates cool, the CLT had to be extended to account for these strains. This was done by using non-linear strain-displacement equations with the Rayleigh-Ritz technique and minimizing the total potential energy. This gave better predictions of the post-cure shape for the thin laminates.

Numerical models were then done to get more accrate predictions of the post-cure room temperature shapes of the thin unsymmetric laminates. Initial pressure was added to the models to simulate the press from the mold. After the initial press was applied, the laminates cooled to room temperature. The numerical models were predicting similar room temperature shapes as the analytical models. After fabricating these laminates, it was observed that the analytical and numerical models were reasonably accurate when predicting the room temperature shapes of the thin laminates. Therefore, the models served as reasonable predictions necessary for piezoelectric actuation. Basic analytical techniques were given for modeling the piezoelectric actuation. Along with experimentation, is was shown that cure process models were able to accurately capture the steps and effects needed later in piezoelectric actuation experimentation.

## REFERENCES

1. P. Giddings, C. R. Bowen, R. Butler and H. A. Kim, "Characterisation of Actuation Properties of Piezoelectric Bi-Stable Carbon-Bre Laminates," Composites, Vol. 39, No. 4, 2008, pp. 697-703.
2. M. L. Dano and M. W. Hyer, "Snap-Through of Unsymmetric Ber-Reinforced Composite Laminates," International Journal of Solids and Structures, Vol. 39, 2001, pp. 175-198.
3. M. L. Dano and M. W. Hyer, "Thermally-Induced Deformation Behavior of Unsymmetric Laminates," International Journal of Solids and Structures, Vol. 35, No. 17, 1998, pp. 2101-2120. doi:10.1016/S0020-7683(97)00167-4
4. C. R. Bowen, R. Butler, R. Jervis, H. A. Kim and A. L. T. Salo, "Morphing and Shape Control Using Unsymmetrical Composites," Intelligent Material Systems and Structures, Vol. 22, No. 18, 2007, pp. 89-98.
5. M. W. Hyer and A. Jilani, "Predicting the Deformation Characteristics of Rectangular Unsymmetric Laminated Piezoelectric Materials," Smart Material Structures, Vol. 7, No. 6, 1998, pp. 784-791. doi:10.1088/0964-1726/7/6/006
6. M. Schlecht, K. Schulte and M. W. Hyer, "Advanced Calculation of the Room-Temperature Shapes of Thin Unsymmetric Composite Laminates," Composite Structures, Vol. 32, No. 1, 1995, pp. 627-633. doi:10.1016/0263-8223(95)00080-1
7. D. N. Bettes, I. T. Salo, C. R. Bowen and H. A. Kim, "Characterization

and Modeling of the Cured Shapes of Arbitrary Layup Bi-Stable Composite Laminates," Composite Structures, Vol. 92, No. 7, 2010, pp. 1694-1700. doi:10.1016/j.compstruct.2009.12.005

8. M. W. Hyer, "Stress Analysis of Ber-Reinforced Composite Materials," McGraw-Hill Companies, New York, 1998.
9. R. M. Jones, "Mechanics of Composite Materials," Taylor and Francis Group, New York, 1999.
10. M. Gigliotti, M. R. Wisnom and K. D. Potter, "Development of Curvature during the Cure of as4/8552 [0/90] Unsymmetric Composite Plates," Composites Science and Technology, Vol. 63, No. 2, 2003, pp. 187-197. doi:10.1016/S0266-3538(02)00195-1
11. F. Mattioni, P. M. Weaver, K. D. Potter and M. I. Friswell, "Analysis of Thermally Induced Multi-Stable Composites," International Journal of Solids and Structures, Vol. 45, No. 2, 2008, pp. 657-675. doi:10.1016/j.ijsolstr.2007.08.031
12. P. Portela, P. Camanho, P. Weaver and I. Bond, "Analysis of Morphing Multi Stable Structures Actuated by Piezoelectric Patches," Computers and Structures, Vol. 86, No. 3-5, 2008, pp. 347-356. doi:10.1016/j.compstruc.2007.01.032
13. C. G. Diaconu, P. M. Weaver and A. F. Arrieta, "Dynamic Analysis of Bi-Stable Composite Plates," Sound and Vibration, Vol. 322, No. 4-5, 2009, pp. 987-1004.doi:10.1016/j.jsv.2008.11.032
14. K. D. Cowley and P. W. R. Beaumont, "The Measurement and Prediction of Residual Stresses in Carbon Ber/ Polymer Composites," Composite Science and Technology, Vol. 57, No. 11, 1997, pp. 1445-1455. doi:10.1016/S0266-3538(97)00048-1
15. L. Ren and A. Parvizi-Majidi, "Cured Shape of Cross-Ply Composite Thin Shells," Composite Materials, Vol. 37, No. 20, 2003, pp. 1801-1820.
16. M. Gigliotti, R. Wisnom and K. D. Potter, "Loss of Bifurcation and Multiple Shapes of Thin [0/90] Unsymmetric Composite Plates Subject to Thermal Stress," Composite Science and Technology, Vol. 64, No. 1, 2004, pp. 109- 128.
17. M. W. Hyer, "Calculations of the Room Temperature Shapes of Unsymmetric Laminates," Composite Materials, Vol. 15, 1981, pp. 296-309.
18. M. R. Wisnom, M. Gigliotti, N. Ersoy, M. Campbell and K. D. Potter, "Mechanisms Generating Residual Stresses and Distortion during Manufacture of Polymer-Matrix Composite Structures," Composites: Part A, Vol. 37, No. 4, 2006, pp. 522-529.doi:10.1016/j.

compositesa.2005.05.019

19. PTC, "Mathcad Users Guide," mathCAD 14.0 Edition, 2007.
20. A. J. Vizzini, "Introduction to Composite Materials," Department of Aerospace Engineering, University of Maryland, College Park, 1990.
21. COMSOL Multi-Physics 3.5a, "Structural Mechanics Module Users Guide," 2008.
22. D. J. Leo, "Engineering Analysis of Smart Material Systems," John Wiley & Sons, Inc., Hoboken, 2007.
23. M. R. Schultz, M. W. Hyer, R. B. Williams, W. K. Wilkie and D. J. Inman, "Snap-Through of Unsymmetric Laminates Using Piezocomposite Actuators," Composite Science and Technology, Vol. 66, No. 14, 2006, pp. 2442- 2448.doi:10.1016/j.compscitech.2006.01.027

# Chapter 8

# THERMO-MECHANICAL PROPERTIES OF SILICON CARBIDE FILLED CHOPPED GLASS FIBER REINFORCED EPOXY COMPOSITES

Gaurav Agarwal[1] Amar Patnaik[2] and Rajesh Kumar Sharma[2]

[1]Department of Mechanical Engineering, S.R.M.S.C.E.T, Bareilly 243122, India

[2]Department of Mechanical Engineering, National Institute of Technology, Hamirpur 177005, India

## ABSTRACT

The effect of addition of silicon carbide (SiC) filler in different weight percentages on physical properties, mechanical properties, and thermal properties of chopped glass fiber-reinforced epoxy composites has been investigated. Physical and mechanical properties, i.e., hardness, tensile strength, flexural strength, interlaminar shear strength, and impact strength, are determined with the change in filler content to notice the behavior of composite material subjected to loading. Thermo-mechanical properties of the material are measured with the help of a dynamic mechanical analyzer. The result shows that the physical and mechanical properties of SiC-filled glass fiber-reinforced epoxy composites are better than unfilled glass fiber-reinforced epoxy composites. Viscoelastic analysis for different compositions indicate that adding too much SiC content results in degradation in energy absorption capacity of the material and hence overall performance of the composites, whereas adding too much (more than 10 wt.%) SiC content increases the elastic behavior of the composite.

## INTRODUCTION

A test for mechanical properties is a prime area of interest. The mechanical properties of a material are those properties that involve a reaction to an applied load. The mechanical properties of a composite determine the range of usefulness of a material and establish the service life that can be expected. Most structural materials are anisotropic, which means that their mechanical properties vary with orientation (Shalin [1995]). The variation in the properties can be due to the change in the microstructure of fiber/filler and matrix reinforcement. The materials can be subjected to many different loading scenarios, and the composite performance is dependent on the loading conditions. The common properties are considered to determine the loading constraints on the durability and service life effect tensile strength, flexural strength, hardness, interlaminar shear strength (ILSS), and impact strength (Rawlings [1999]; Reid and Zhou [2000]).

It is also necessary to determine the effect of change in temperature on the properties of fiber-reinforced epoxy composites because an increase in temperature results in the decrease in strength and ductility, whereas a decrease in temperature results in the increase in ductility and strength (Nielsen and Landel [1994]). Apart from fiber-reinforced polymer composites, the composites made from both fiber/filler reinforcement performed well in many practical situations. Silicon carbide (SiC) filler particles were added to fiber-reinforced epoxy to enhance the mechanical properties of composites. A change in particle size, shape, and percentage content on the mechanical properties of fiber-reinforced polymer composites is an area of keen interest (Koh et al. [1993]; Imanaka et al. [2001];

Yamamoto et al. [2003]). From research, it has been found that the structure and shape of silica particles have significant effects on the mechanical properties such as fatigue resistance, tensile, and fracture properties (Nakamura et al. [1991a], [b]; Nakamura et al. [1992]).

The literature mentioned above gives a brief overview about the effects of addition of fiber/filler to mechanical and thermal properties of SiC-filled glass fiber-reinforced epoxy composites. It has been observed that various mechanical properties at optimum fiber loading conditions can be determined with the addition of fiber/filler reinforcement on matrix composites. To the author's knowledge, no work has been done to determine the optimum composition with the combination of fiber/filler and matrix to optimize the performance of the composite material.

## MATERIALS AND METHODS

### Material required

Chopped E-glass fiber (5 to 10 mm long, 200 G.S.M. density) manufactured by Ciba Geigy and locally supplied by Northern Polymers Ltd., New Delhi, India, is used as a fiber reinforcement. Commercially available SiC powder also known as carborundum of particle size 25 to 60 μm (density 2.6 g/cm$^3$) obtained from Silcarb Recrystalized Pvt. Ltd., Bangalore, India, is used as a particulate filler. The matrix material consists of epoxy resin LY 556 and room-temperature curing hardener HY951 supplied by Crystal Chemicals, Delhi, India. The composites were fabricated by blending epoxy resin, glass fiber, and SiC filler in a certain weight percentage reinforcement. Five different compositions of composites were prepared by varying the SiC filler reinforcement with fixed weight percentage (wt.%) of chopped glass fiber reinforcement. SiC filler in five different weight percentages (0, 5, 10, 15, and 20 wt.%) are added with fixed 20 wt.% of chopped glass fiber and remaining epoxy so as to notice the effect of SiC reinforcement on physical, mechanical, and thermal properties of chopped glass fiber-reinforced epoxy composites. The composites were prepared by blending certain weight percentage of fiber/filler and epoxy resin in separate containers and then poured in a mold of desired dimensions. Labeling was done with the help of rollers, and suitable weights were applied on top of the mold. Similarly, five different compositions of fiber, filler, and epoxy resin are poured in separate molds by varying five percentages of SiC filler content with that of the epoxy resin keeping the chopped glass fiber weight percentage as constant. The composites are then left for solidification at room temperature for 24 h. After the solidification process, the composites are then removed from the mold and marking is done as per the test standards. Specimens were prepared as per American Society for Testing Materials (ASTM) test standards for tensile, flexural, interlaminar shear strength, impact strength, and hardness tests.

## Tests for mechanical properties

The mechanical properties were determined to understand the behavior of the material under different loading conditions. To understand how materials deform or break as a function of applied load, time, temperature, and other conditions, tensile strength, impact strength, flexural strength, interlaminar shear strength, and hardness of the composite specimens were determined.

The capability of the composite material to resist abrasion or indentation is known as hardness. A Rockwell hardness tester is used to measure the hardness of the composite specimen. A spherical indenter (1/16 in or 1.5875 mm diameter) was forced into the surface of the material under conditions of controlled magnitude and rate of loading. The size of indentation is the measure of amount of hardness of the specimen.

Tests for mechanical properties, i.e., tensile test, interlaminar shear strength, and flexural strength, are done on a universal testing machine (UTM) Instron 1195. The tensile test is performed to measure the ductility of the material. Equal and opposite loads are applied at both ends of the rectangular specimen on the UTM Instron 1195 (ASTM [1976]). Dimensions of the test specimen are 200 mm long, 11.5 mm wide, and with variable thickness (depending on the composition of the composite material). Flexural strength is the ability of the material to resist bending under the application of load. A three-point bend test was performed on the same UTM Instron 1195 with a span length of 30 mm in between the supports and a crosshead speed of 10 mm/min to measure the amount of failure when subjected to loading. Rectangular test specimen with dimensions 157 mm length, 12.7 mm width, and 6.35 mm thickness has been used for the experiment. Interlaminar shear strength tests are conducted as per ASTM D 2344-84 test standards on UTM Instron 1195 with span length of 50 mm and crosshead speed of 10 mm/min is maintained for testing (ASTM [1984]).

Impact tests are done as per ASTM D 256 test standards (ASTM [1997]) on plastic/composite specimen digital impact testing machine. Impact test is done to notice the energy consuming capacity of the composite material before fracture. Charpy V-notch impact test samples with dimensions 64 mm × 12.7 mm × 3.2 mm were prepared with a V-notch groove (2.5 mm depth) in the center of the specimen. The specimen was fixed on the impact tester such that the notch was at the opposite face to the striking end of the hammer.

## RESULTS AND DISCUSSION

The physical and mechanical properties describe the behavior of the material to various practical applications. The properties such as hardness describe the physical state of the system. The mechanical property of the material is a measure of the behavior of the material under different loading conditions. Tests were done to notice the effect of addition of SiC filler on the physical and mechanical properties and the optimum fiber/filler loading of glass fiber-reinforced epoxy composites at which specific wear rate in minimum.

### Effect of hardness on SiC-filled chopped glass fiber-reinforced epoxy composites

Hardness is a surface property and is a measure of wear resistance on the surface of the composite. The measured hardness values of all the five weight percentages of SiC reinforcement are shown in Figure 1. It can be seen that the hardness increases with the increase in the SiC content at a constant rate. Hardness increases linearly from 0 wt.% SiC content to 15 wt.% SiC content, but further increase in SiC content results in the decrease in the value of hardness. The increase in the value of hardness are attributed to the fact that as the density of composition increases due to filler particles introduced between the fiber and the matrix, the compression value of the specimen decreases and hardness increases. Addition of a small volume fraction of graphite, SiC, and alumina can significantly improve the hardness and wear resistance of the composites. With the addition of SiC filler content, the filler particles (SiC) fill in the gap between the fiber and the matrix and form a more dense structure and hence hardness increases (Mohamed et al. [2009]).

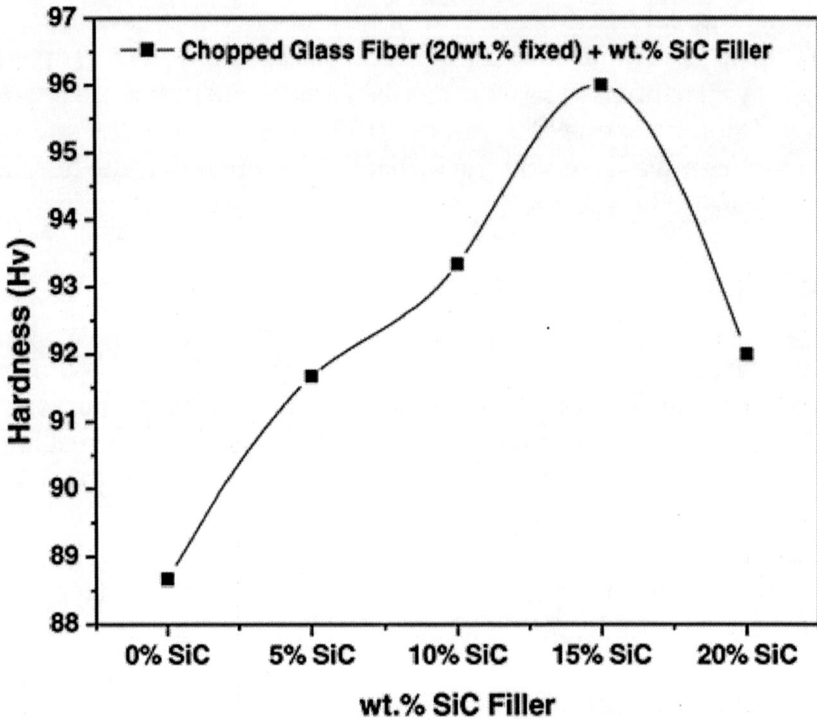


**Figure 1. Effect of hardness on SiC-filled chopped glass fiber-reinforced epoxy composites.**

## Effect of tensile strength on SiC-filled chopped glass fiber-reinforced epoxy composites

The ultimate tensile strength is an engineering value calculated by dividing the maximum load on a material by the initial cross section of the test specimen. Figure 2 shows the graph of tensile strength versus the SiC-filled chopped glass fiber-reinforced epoxy composite. Tensile strength increases from 0 to 10 wt.% SiC content, whereas further increase in SiC content results in the decrease in the value of tensile strength. This is because filler particles act as a barrier in transferring stress from one point to another, and an increase in SiC content beyond 10 wt.% results in the increase of transfer of stresses from one point to another. Also as the fiber/filler content increases, the bonding surface area increases and hence bonding strength decreases. Due to insufficient amount of bonding between three different constituents, the loads may not effectively be transferred from one end to another and hence there is reduction in tensile strength of the composite. Tensile strength of carbon fiber-reinforced polyetheramide improves significantly with the addition in the percentage of fiber/filler reinforcement. The extent of improvement of tensile strength however is not proportional to the amount of carbon fiber reinforcement (Sua et al. [2005]; Devendra and Rangaswamy [2012]).

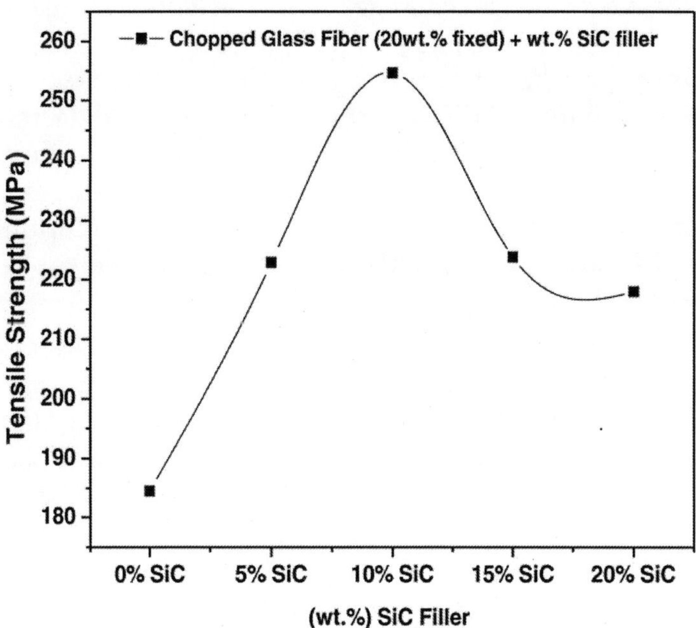


**Figure 2.** Effect of tensile strength on SiC-filled chopped glass fiber-reinforced epoxy composites.

## Effect of flexural strength and interlaminar shear strength on SiC-filled chopped glass fiber-reinforced epoxy composites

Figure 3 shows the flexural strength of SiC-filled glass fiber-reinforced epoxy composites. Flexural strength increases from 55 to 87 MPa with the increase in SiC filler content from 0 to 10 wt.%; an increase in the flexural strength beyond 10 wt.% SiC content results in the decrease in the value of SiC content. When the specimen is placed on two support points and load is applied from the top of the specimen, then the specimen is subjected to bending and the top layer is subjected to compressive loading, whereas the bottom layer is subjected to tensile loading.

When the bonding between the fiber/filler and the matrix is increased, a flexural strength increases and strong bonding transfers loads from one end to another resulting in the increase in flexural strength of the specimen, whereas when the percentage of fiber/filler exceeds the required percentage, then the surface area increases, while the weight percentage of the matrix decreases; therefore, the bonding strength decreases, and loads cannot be effectively transferred from one part to another resulting in the decrease in the flexural strength of the composite. Another reason for the reduction in flexural strength with the addition of SiC filler is that the addition of SiC content beyond 10 wt.% disturbs the matrix continuity and reduction in bonding strength between filler, matrix, and fiber (Kaundal et al. [2012]).

Interlaminar shear strength is the measure of shear strength between the layers of fibers. Figure 4 shows the ILSS value of SiC-filled glass fiber-reinforced epoxy composites. The ILSS value increases with the increase in the value of SiC content because SiC filler particles stick to the ends of fibers and act as a barrier in the transfer of shear stress from one part to another. A proportional graph is obtained between the weight percentage of SiC filler reinforcement and ILSS value, i.e., ILSS increases in equal proportion with the increase in the value of SiC content.

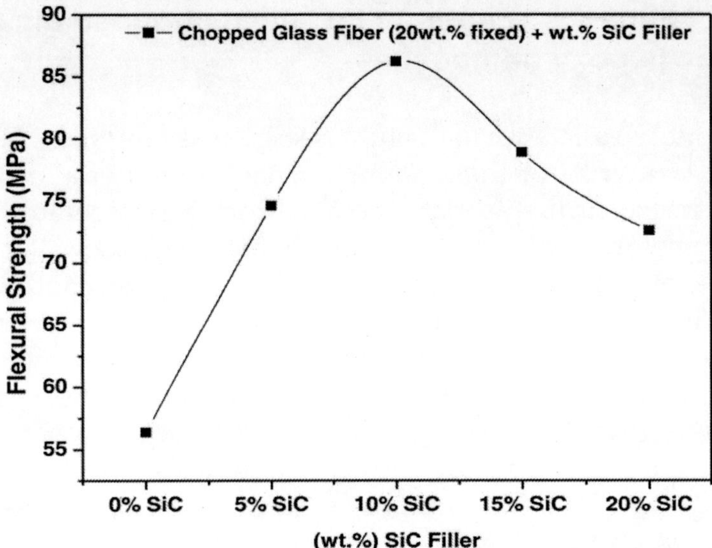


**Figure 3.** Effect of flexural strength on SiC-filled chopped glass fiber-reinforced epoxy composites.

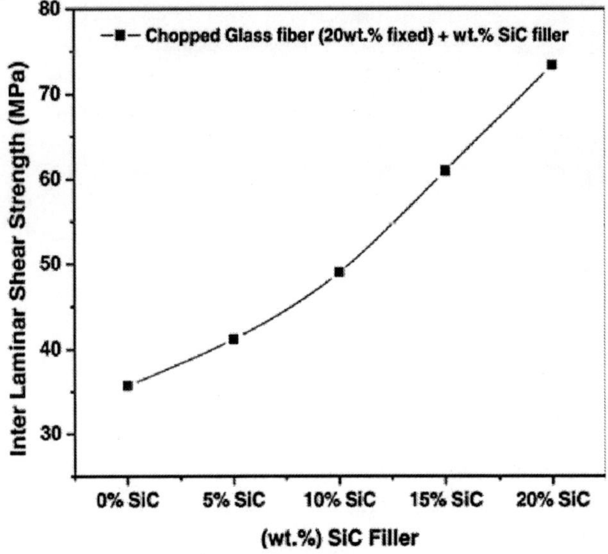


**Figure 4.** Effect of ILSS on SiC-filled chopped glass fiber-reinforced epoxy composites.

## Effect of impact strength on SiC-filled chopped glass fiber reinforced epoxy composites

The material's resistance to fracture is known as toughness. It is often expressed in terms of the amount of energy a material can absorb before fracture. A ductile material can absorb a considerable amount of energy before fracture, while a brittle material absorbs very little. Figure 5shows the increase in the value of impact strength with the increase in the percentage of fiber/filler reinforcement. The value of impact strength increases from 0 to 15 wt.% and further decreases for 20 wt.% SiC filler reinforcement. The increase in impact strength is due to the weaker bond, i.e., as the bond strength between the glass fiber, SiC filler, and the epoxy resin reduces, the impact energy absorbing capacity of the composite increases. Also, a large amount of energy is absorbed by the crack initiated along the fiber/filler/matrix interface during debonding. The energy absorbing capability of composites depends on the properties of the constituents. Based on the literature review, it has been found that the benefits of using SiC as reinforcement are improved stiffness, strength, and chemical stability (Kaundal et al. [2012]).

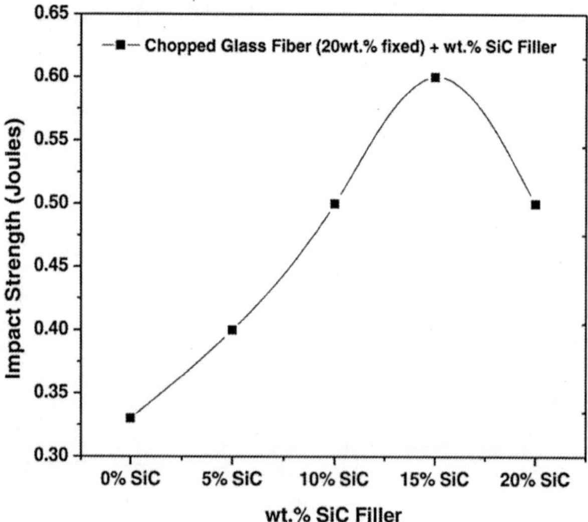


**Figure 5.** Effect of impact strength on SiC-filled chopped glass fiber-reinforced epoxy composites.

## Thermo-mechanical properties of composite (dynamic mechanical analysis)

A common cause of tribological failure of composite parts is the inability to support a given load due to the loss in modulus at frictionally induced elevated temperatures. Dynamic mechanical analysis (DMA) has emerged as one of the most powerful tools available for the study of the behavior of materials. Simply stated, DMA measures the viscoelastic properties of materials. DMA plot provides values of those temperature regions where material properties are very stable with temperature and where rapid changes will occur that would render the material useless (Ward and Sweeney [2004]). One important application of DMA is measurement of the glass transition temperature of polymers. Amorphous polymers have different glass transition temperatures, above which the material will have rubbery properties instead of glassy behavior and the stiffness of the material will drop dramatically with an increase in viscosity. At the glass transition, the storage modulus decreases dramatically and the loss modulus reaches a maximum (Agarwal et al.[2013]). DMA of chopped glass fiber SiC filler-reinforced epoxy composite have been carried out in this study to investigate the variation of storage modulus (E'), loss modulus (E") and damping factor (tan delta) i.e.:

$$\tan\delta = \frac{E'}{E''} = \frac{\text{storage modulus}}{\text{loss modulus}} \qquad (1)$$

Equation 1 represents that tan $\delta$ is the ration of storage modulus to loss modulus, i.e., as the value of loss modulus increases, the value of storage modulus decreases (Agarwal et al. [2013]). Similar trend is noticed in Figures 6 and 7. All tests run as per ASTM D4065-94 test standards using a fixed frequency of oscillation of 1 Hz and a sample heating rate of 2°C/min. All tests were initiated from room temperature and goes to 200°C temperature. The mode of stress applied is flexure, and the fixture configuration is a single cantilever beam. Figures 6, 7, and 8 show the most common graphic representation for elastic or storage modulus, the viscous or loss modulus, and tan delta as a function of temperature, respectively.

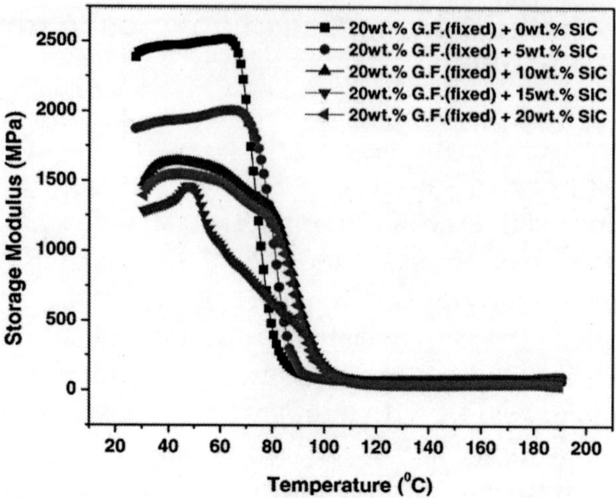


**Figure 6.** Variation of storage modulus for SiC filled chopped glass fiber-reinforced epoxy composites.

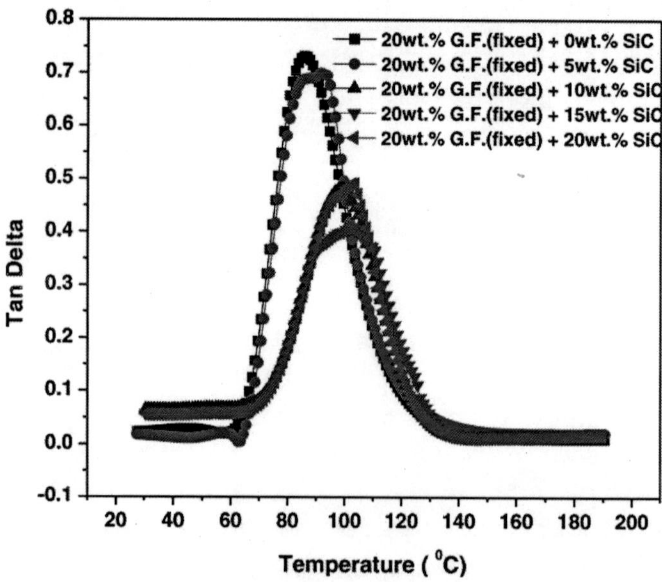


**Figure 7.** Variation of loss modulus for SiC filled chopped glass fiber-reinforced epoxy composites.

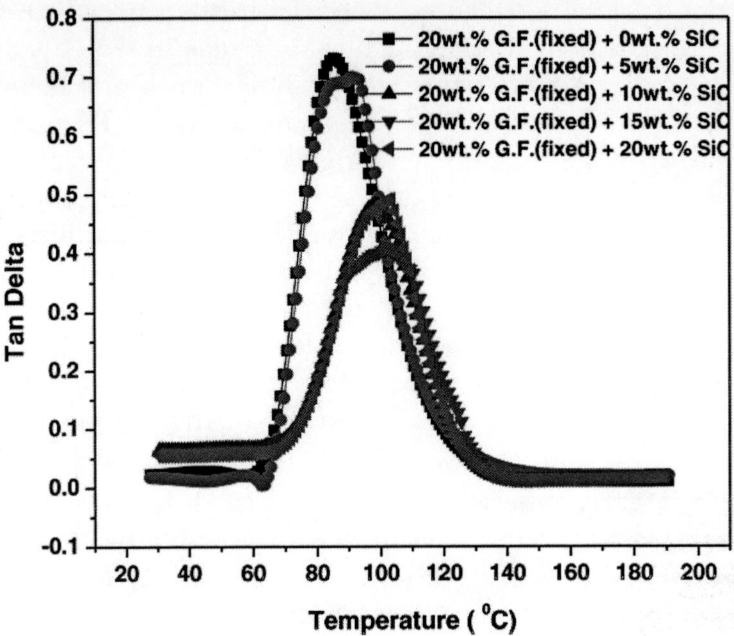


**Figure 8.** Variation of tan delta for SiC filled chopped glass fiber-reinforced epoxy composites.

Figure 6 shows a graph of storage modulus versus temperature for SiC-filled chopped glass fiber-reinforced epoxy composite. Storage modulus values for chopped glass fiber SiC filler-reinforced epoxy composites show a linear trend with slight decrease in the value of storage modulus with the change in temperature (room temperature to 70°C). This shows glassy regime in region 1, i.e., from room temperature to 65°C. Region 2 shows a sharp decline in the values of storage modulus with the increase in the values of temperature (65°C to 90°C), i.e., the material changes from glassy to rubbery transition regime in the temperature range of 65°C to 90°C; due to a sharp drop in the value of storage modulus, the material has lost its usefulness as a structural material. Due to amorphous structure in the polymer and the presence of unorganized intermolecular structure, the storage modulus value drops suddenly. Region 3 again is a straight line showing negligible decrease in the values of storage modulus with the increase in the value of temperature. Here, the values of storage modulus nearly tend to zero, which shows a rubbery regime indicating degradation of the moduli above 90°C.

In the glassy state, i.e., at low temperatures, stiffness is related to the small displacement of molecules, whereas in the rubbery state at high temperature, molecular chains have considerable flexibility; so that in the undeformed state, they can adopt conformations that lead to maximum entropy. The higher values in region 1 and sharp decrease in the values in region 2 are due to the fact that in region 1, the material is in glassy state in which the contribution of elastic modulus is more than the viscous modulus, whereas in region 2, the material is in glass transition stage in which a change from glass transition state into rubber elastic state takes place. In the glass transition region, i.e., from glassy to the rubbery transition regime region, the value of storage modulus falls during heating to a level of 1,000th to 10,000th of its original value (Jinmin et al. [1994]).

Figure 7 clearly shows the value of loss modulus with the increase in the value of temperature. From the values obtained in the graph, we notice that the values of loss modulus rise to a maximum as the storage modulus values are in their most rapid rate of decent. The peak of the loss modulus curve denotes the glass transition temperature ($Tg$). Figure 7 denotes the glass transition temperature as 90°C and desired loss modulus values (minimum values) obtained for 10 to 20 wt.% SiC content. This may be due to the increase in the percentage of fiber/filler content, i.e., as the percentage of fiber/filler content increases, the height of the E″ curve decreases and the peak width spreads across a wider temperature range leading to a decrease in the value of loss modulus (desired value obtained). Further, it has also been observed from Figure 7 that the composition with lower fiber/filler content peak of glass transition temperature shifted towards lower temperature with a sharp peak and higher value of loss modulus.

Figure 8 shows the graph of tan delta versus temperature. The tan delta curve closely resembles the loss modulus curve leading to the glass transition tan delta values well below 0.1. The rapid rise in tan delta values closely resembles the rapid decline in storage modulus values. The peak value of tan delta indicates that the material is nonelastic, whereas the lower value of tan delta indicates that the material is elastic in nature (Agarwal et al. [2013]). Region 1 shows that the material is highly elastic in this region, and as the value of tan delta shifts to region 2, the nature of material shifts from elastic to plastic. The peak value of tan delta indicates that any further increase in temperature beyond this changes the material from elastic to plastic zone. Figure 8shows that as the percentage of fiber/filler increases, the peak value shifts towards higher temperature resulting in wider elastic range of temperature, whereas as the percentage of fiber/filler decreases, the peak value shifts towards the lower temperature range resulting in narrow elastic range.

## COLE-COLE PLOT ANALYSIS

Figure 9 shows the analysis carried on SiC-filled chopped glass fiber-reinforced epoxy composites using Cole-Cole analysis. A Cole-Cole plot is used to predict the nature of a composite whether it be homogeneous or heterogeneous (Kumar et al. [2011]). A Cole-Cole is plotted for storage modulus and loss modulus where the values of loss modulus (log E″) are plotted on $x$-axis and the values of storage modulus (log E′) are plotted on $y$-axis. Cole-Cole plot is used to describe the analysis of molecular architecture. Changes in the architecture of the model samples were readily detected as systematic variations in shape and displacements on Cole-Cole plot. A homogeneous system typically exhibits a semicircular curve, whereas a heterogeneous system typically exhibits an irregular curve (imperfect semicircle); also, the nature of curvature obtained in Cole-Cole curve determines the adhesion between fiber and matrix.

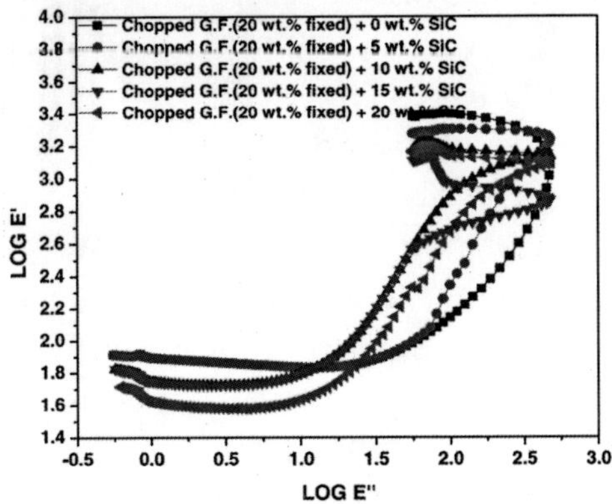


**Figure 9.** Cole-Cole plot for SiC-filled chopped glass fiber-reinforced epoxy composites

Investigated values of SiC-filled chopped glass fiber-reinforced epoxy composites represent imperfect semicircles and hence are heterogeneous in nature. Also, unfilled glass fiber-reinforced epoxy composites exhibit better interfacial characteristics and some homogeneous nature with respect to SiC-filled glass fiber-reinforced epoxy composites. This is because adding SiC content breaks the matrix continuity, and the nature is shifted more towards heterogeneous tendency.

## CONCLUSIONS

Experimental observations are carried out on SiC-filled chopped glass fiber-reinforced epoxy matrix composite to notice the effect of physical, mechanical, and thermal properties. Based on the experimental observations, the following conclusions can be drawn:

1. Mechanical properties such as hardness, tensile strength, interlaminar shear strength, flexural strength, and impact strength increase with the increase in SiC filler content up to 10 to 15 wt.%, whereas addition of SiC content beyond 15 wt.% results in the decrease in mechanical property.
2. The value of hardness and impact strength increases up to 15 wt.% filler content; the value of tensile strength and flexural strength increases up to 10 wt.% SiC filler content, whereas an exceptional increase in the value of ILSS is noticed up to 20 wt.% SiC filler content.
3. Storage modulus values for SiC-filled glass fiber-reinforced epoxy composite decreases with the increase in SiC content. Storage modulus values decrease with the increase in the percentage of SiC content, whereas elastic range of the material increases with the increase in SiC content.
4. As the percentage of SiC content increases, the area under the curve spreads across the wider range and the peak value shifts towards the higher temperature region resulting in the increase in elastic range of the composite.
5. SiC-filled chopped glass fiber-reinforced epoxy composites represent imperfect semicircles and hence are heterogeneous in nature. Also, unfilled glass fiber-reinforced epoxy composites exhibit better interfacial characteristics and some homogeneous

nature with respect to SiC-filled glass fiber-reinforced epoxy composites. The imperfect circles and heterogeneous nature is due to the uneven distribution of SiC particles inside the composite.

6. From the analysis of the results of mechanical properties and thermal properties, it has been concluded that the optimum properties are obtained for 15 wt.% SiC content in addition to 20 wt.% chopped glass fiber-reinforced epoxy composite.

### Competing interests

The authors declare that they have no competing interests.

## AUTHORS' CONTRIBUTIONS

All authors read and approved the final manuscript.

## REFERENCES

1. Agarwal G, Patnaik A, Sharma RK: **Parametric optimization of three-body abrasive wear behavior of long and short carbon fibre reinforced epoxy composites.**
2. *Tribology Material, surface and interfaces* 2013, **7**(3):150-160.
3. ASTM (American Society for Testing and Materials): *ASTM D 3039-3076: tensile properties of fiber–resin composites*. West Conshohocken: American National Standard; 1976.
4. ASTM (American Society for Testing and Materials): *ASTM D 2344-84: standard test method for apparent inter-laminar shear strength for parallel fiber composites by short beam method. Annual book of ASTM standards.* West Conshohocken: ASTM; 1984:15-17.
5. ASTM (American Society for Testing and Materials) (1997) Standard D 256–97: standard test methods for determining the pendulum impact resistance of notched specimens of plastic: *Annual book of ASTM standards, 8(1)*. West Conshohocken: ASTM; 1999:01-20.
6. Devendra K, Rangaswamy T: **Determination of mechanical properties of $Al_2O_3$, Mg $(OH)_2$ and SiC filled E-glass/epoxy composites.**
7. *Int J Eng Res Appl* 2012, **2**(5):2028-2033.
8. Imanaka M, Takeuchi Y, Nakamura Y, Nishimura A, Lida T: **Fracture**

**toughness of spherical silica-filled epoxy adhesives.**

9. *Int J Adhesin Adhes* 2001, **21:**389-396.
10. Zhang J, Perez RJ, Wong CR, Lavernia EJ: **Effects of secondary phases on the damping behavior of metals, alloys and metal matrix composites.**
11. *Mater Sci Eng* 1994, **13**(8):325-389.
12. Kaundal R, Patnaik A, Satapathy A: **Comparison of the mechanical and thermo-mechanical properties of unfilled and SiC filled short glass polyester composites.**
13. *Silicon* 2012, **4:**175-188.
14. Koh SW, Kim JK, Mai YW: **Fracture toughness and failure mechanisms in silica-filled epoxy resin composites: effects of temperature and loading rate.**
15. *J Polymer* 1993, **34**(16):3446-3455.
16. Kumar S, Satapathy BK, Patnaik A: **Visco-elastic interpretations of erosion performance of short aramid fibre reinforced vinyl ester resin composites.**
17. *J Mater Sci* 2011, **46:**7489-7500.
18. Mohamed SS, Mahmoud TS, Eimahallawi IS, Khalifa TA: **Mechanical and tribological characteristics of epoxy based PMC's reinforced with different ceramic particulates.**
19. *Eng Res J* 2009, **121:**29-49.
20. Nakamura Y, Yamaguchi M, Kitayama A, Okubo M, Matsumoto T: **Effect of particle size on fracture toughness of epoxy resin filled with angular-shaped silica.**
21. *J Polymer* 1991, **32**(12):2221-2229.
22. Nakamura Y, Yamaguchi M, Okubo M, Matsumoto T: **Effect of particle size on impact properties of epoxy resin filled with angular shaped silica particles.**
23. *J Polymer* 1991, **32**(16):2976-2979.
24. Nakamura Y, Yamaguchi M, Okubo M, Matsumoto T: **Effects of particle size on mechanical and impact properties of epoxy resin filled with spherical silica J.**
25. *Appl Polym Sci* 1992, **45:**1281-1289.
26. Nielsen LE, Landel RF: *Mechanical properties of polymers and composites.* 2nd edition. New York: Marcel Deckker; 1994.
27. Rawlings MR: *Composite materials: engineering and science.* Cambridge: Woodhead Publishing; 1999.

28. Reid SR, Zhou G: *Impact behavior of fiber reinforced composite materials and structures*. London: Taylor & Francis; 2000.

29. Shalin RE: *Polymer matrix composites. Soviet advanced composites technology series*. 4th edition. Heidelberg: Springer; 1995.

30. Sua F, Zhang Z, Wang K, Jiang W, Liu W: **Tribological and mechanical properties of the composites made of carbon fabrics modified with various methods.**

31. *Compos Part A* 2005, **36:**1601-1607.

32. Ward IM, Sweeney J: *An introduction to the mechanical properties of solid polymers*. New York: Wiley; 2004.

33. Yamamoto I, Higashihara T, Kobayashi T: **Effect of silica-particle characteristics on impact/usual fatigue properties and evaluation of mechanical characteristics of silica-particle epoxy resins.**

# Chapter 9

# OIL PALM BIOMASS FIBERS AND RECENT ADVANCEMENT IN OIL PALM BIOMASS FIBERS BASED HYBRID BIO COMPOSITES

H.P.S. Abdul Khalil[1], M. Jawaid[2], A. Hassan[2], M.T. Paridah[3] and A. Zaidon[4]

[1] School of Industrial Technology, Universiti Sains Malaysia, Penang, Malaysia

[2] Department of Polymer Engineering, Faculty of Chemical Engineering, Universiti Teknologi Malaysia, Skudai, Johor, Malaysia

[3] Institute of Tropical Forestry and Forest Products, Universiti Putra Malaysia, Serdang, Selangor Darul Ehsan, Malaysia

[4] Faculty of Forestry, Universiti Putra Malaysia, Serdang, Selangor Darul Ehsan, Malaysia

## INTRODUCTION

Worldwide 42 countries cultivate *Elaeis guineensis* (oil palm tree) on about 27 million acres. Oil palm is one of the most valuable plants in Malaysia, Indonesia and Thailand. Oil palm tree (Figure 1)generally has an economic life span of about 25 years, and it contributes to a high amount of agricultural waste in Malaysia. The oil palm tree is ≈ 7–13m in height and 45–65 cm in diameter, measuring 1.5m above the ground level (Abdul Khalil et al. 2010d) and one of the commercial crop in Malaysia. Malaysia is the world's largest producer and exporter of the oil palm, accounting for approximately 60% of the world's oil and fat production. The oil palm industry in Malaysia, with its 6 million hectares of plantation, produced over 11.9 million tons of oil and 100 million tons of biomass (Abdul Khalil et al. 2010b). The amount of biomass produced by an oil palm tree, inclusive of the oil and lignocellulosic materials, is on the average of 231.5 kg dry weight/year (Abdul Khalil et al. 2010c). An estimation based on a planted area of 4.69 million ha (MPOB 2009) and a production rate of dry oil palm biomass of 20.34 tonnes per ha per year (Lim 1998) show that the Malaysian palm oil industry produced approximately 95.3 million tonnes of dry lignocellulosic biomass in 2009. This figure expected to increase substantially when the total planted hectarage of oil palm in Malaysia could reach 4.74 million ha in 2015 (Basiron and Simeh 2005), while the projected hectarage in Indonesia is 4.5 million ha.Oil palm production has nearly doubled in the last decade, and oil palm has been the world's foremost fruit crop, in terms of production, for almost 20 years (Abdul Khalil et al. 2010c). Oil palm industries generate abundant amount of biomass say in millions of tons per year (Rozman et al. 2005) which when properly used will not only be able to solve the disposal problem but also can create value added products from this biomass.

OPB is an agricultural by-product periodically left in the field during the replanting, pruning, and milling processes of oil palm. Oil palm biomass (OPB) is classified as lignocellulosic residues that typically contain 50% cellulose, 25% hemicellulose, and 25% lignin in their cell wall(Alam et al. 2009).The biomass from oil palm residue include the oil palm trunk (OPT), oil palm frond (OPF), kernel shell, empty fruit bunch (EFB), presses fruit fibre (PFF), and palm oil mill effluent (POME). Oil palm fronds accounts for 70% of the total oil palm biomass produced, while the EFB accounts for 10% and OPT accounts for only about 5% of the total biomass produced (Ratnasingam 2011). They also stated that 89% of the total oil palm biomass produced annually used as fuel, mulch and fertilizer. In 2006, Malaysia alone produced about 70 million tonnes of oil palm biomass, including trunks, fronds, and empty fruit bunches (Yacob

2007). Despite this enormous production, oil comprises only a small fraction of the total biomass produced by the plantation. The remaining biomass is an immense amount of lignocellulosic materials in the form of fronds, trunks and empty fruit bunch. As such, the oil palm industry must be prepared to take advantage of the situation and utilize the available biomass in the best possible manner (Basiron 2007). Oil palm biomass waste can create substantial environmental problems when simply left on the plantation fields. Presently, EFB mainly used as mulch, but the economic are marginal due to the high transport cost. It is seldom burnt as fuel, as the shell and fruit fibres are sufficient for oil palm mills (Abdul Khalil 2004). It reported that oil palm biomass burnt as fuel in the boiler to produce steam for electricity generation in the processing of oil palm (Nasrin et al. 2008). Researchers stated that a large amount of oil palm residues resulting from the harvest can be utilized as by–products, and it can also help to reduce environmental hazards (Sulaiman et al. 2011). Researchers carried out an extensive study on utilization of OPB as a source of renewable materials (Sumathi et al. 2008). Oil palm biomass fibres offer excellent specific properties and have potential as outstanding reinforcing fillers in the matrix and can be used as an alternative material for bio-composites, hybrid composites, pulp, and paper industries (Abdul Khalil et al. 2010d; Abdul Khalil et al. 2009).

Natural fibres such as hemp, kenaf, jute, sisal, banana, flax, oil palm etc. have been in considerable demand in recent years due to their eco-friendly and renewable nature. Natural fibres received considerable attention as potential reinforcements in polymer composites (Wong et al. 2010; Wan Nadirah et al. 2011; Bledzki and Gassan 1999). The attraction towards utilization of natural fibres as a reinforcement of polymer-based composites is mainly due to their various advantages over synthetic fibres such as are low density, lower cost, light weight, high strength to weight ratio, biodegradability, acceptable specific properties, better thermal and insulating properties (Rout et al. 2001; Rana et al. 2003; Joshi et al. 2004; Nayak et al. 2009). Natural fibre are also less wear and tear in processing, lower energy requirements for processing, wide availability and relative non abrasiveness over traditional reinforcing fibres such as glass and carbon. Natural fibre based polymer composites made of jute, oil palm, flex, hemp, kenaf have a low market cost, attractive with respect to global sustainability and find increasing commercial use in different applications (Jawaid et al. 2011b). Despite the advantages, use of natural fibre reinforced composites has been restricted due to its high moisture absorption tendency, poor wettability, and low thermal stability during processing and poor adhesion with the synthetic counterparts (Demir et al. 2006; Son et al. 2001). Natural fibres are not suitable for high performance military and aerospace applications due to its low strength, environmental sensitivity, and poor moisture resistance which results in degradation in strength and stiffness of natural fibre reinforced composites. Most of the drawbacks that have been identified can be overcome by effective hybridization of natural fibre with synthetic fibre or natural fibre.

**Figure 1.** Oil palm Tree

Hybrid composites are these systems in which one kind of reinforcing material incorporated in a mixture of different matrices (blends) (Thwe and Liao 2003), or two or more reinforcing and filling materials are present in a single matrix (Karger-Kocsis 2000; Fu et al.

2002) or both approaches are combined. Hybrid composite which contain two or more types of fibre in a single matrix, the advantages of one type of fibre could complement with what are lacking in the other. Hybrid composites fabrication by proper material design could help in achieve balance in cost and performance (John and Thomas 2008). Various researchers have tried blending of two fibres in order to achieve the best utilization of the positive attributes of one fibre and to reduce its negative attributes as far as practicable(Abdul Khalil et al. 2009; Abu Bakar et al. 2005; Jawaid et al. 2012;Jacob et al. 2004a; Akil et al. 2009). One another reasons for blending of one fibre with other natural fibres are to impart fancy effect, reduce cost of the end product, and find out suitable admixture of natural origin to mitigate the gap between demand and supply (Basu and Roy 2007). It is possible to combine two or more existing materials and allow a superposition of their properties—in short, to create a *hybrid* (Figure 2)(Ashby and Brechet 2003). Hybrid composites reinforced with natural fibres, well often combined with synthetic fibres such as glass/Carbon fibres, can demonstrate exemplary mechanical performance (Abu Bakar et al. 2005; Wan Busu et al. 2010; Noorunnisa Khanam et al. 2010). Sisal/oil palm fibres and jute/oil palm fibres appear to be promising materials because of the high tensile strength of sisal and jute fibres and the toughness of oil palm fibre (Jacob et al. 2004b; Jawaid et al. 2010). Therefore, any composite comprised of these two fibres will exhibit the desirable properties of the individual constituents. The primary advantages of using oil palm fibres in hybrid composites are its low densities, non abrasiveness and biodegradability. Mixing natural fibres like hemp and kenaf with thermoplastics put Flex Form Technologies (Jon Fox-Rubin 2010) on the map and in the door panels of Chrysler's Sebring convertible. However, the combination of rising oil prices and exterior applications could drive its utilization even higher. Flex Form is also looking to produce vehicle load floors, headliners, seatbacks, instrument panel top covers, knee bolsters, and trunk liners.

## OIL PALM FIBERS

Oil palm industries generates massive quantities of oil palm biomass such as oil palm trunk (OPT), oil palm frond (OPF) and oil palm empty fruit bunch (EFB) as shown in Figure 3. The OPF and OPT generated from oil palm plantation while the oil palm EFB from oil palm processing. In Malaysia, oil palm EFB is one of the biomass materials, which is a by-product from the palm oil industry. EFB are left behind after the fruit of the oil palm harvested for the oil refining process. EFB amounting to 12.4 million tonnes/year (fresh weight) are regularly discharged from palm oil refineries (Abdul Khalil et al. 2010c).

This oil palm EFB has high cellulose content and has potential as natural fiber resources, but their applications account for a small % of the total biomass productions. Several studies showed that oil palm fibres have the potential to be an effective reinforcement in thermoplastics and thermosetting materials (Khalil et al. 2008; Hassan et al. 2010; Shinoj et al. 2011). In order to develop other applications for oil palm fibres they need to be extracted from the waste using a retting process (Shuit2009). Oil palm frond (OPF) is one of the most abundant by-products of oil palm plantation in Malaysia. Oil palm fronds are available daily throughout the year when the palms are pruned during the harvesting of fresh fruit bunches for the production of oil. OPF contains carbohydrates as well as lignocellulose and it amounting to 24 million tons/year discharged from oil palm mills. Oil palm frond, consisting of leaflets and petioles, is a by-product of the oil palm industry in Malaysia and their abundance has resulted in major interest in their potential use for livestock feed (Dahlan 2000). OPF are left rotting between the rows of palm trees, mainly for soil conservation, erosion control and ultimately the long-term benefit of nutrient recycling (Abu Hassan 1994). The large quantity of fronds produced by a plantation each year makes these a very promising source of roughage feed for ruminants.

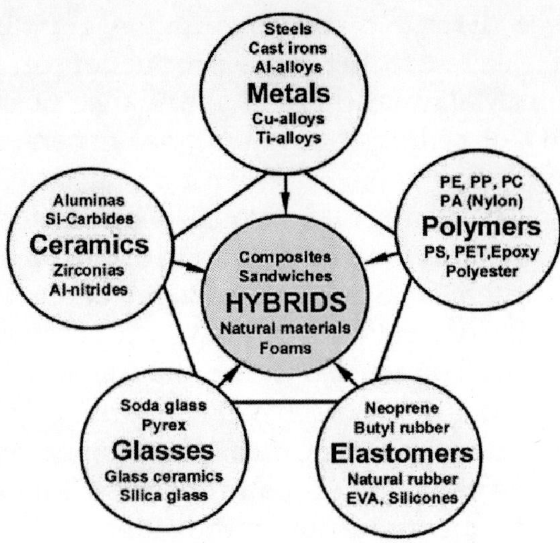


**Figure 2.** Hybrid materials combine the properties of two (or more) monolithic materials, or of one material and space. They include fibrous and particulate composites, foams and lattices, sandwiches and almost all natural materials. One might imagine two further dimension: those of shape and scale (Ashby and Brechet 2003 with permission).

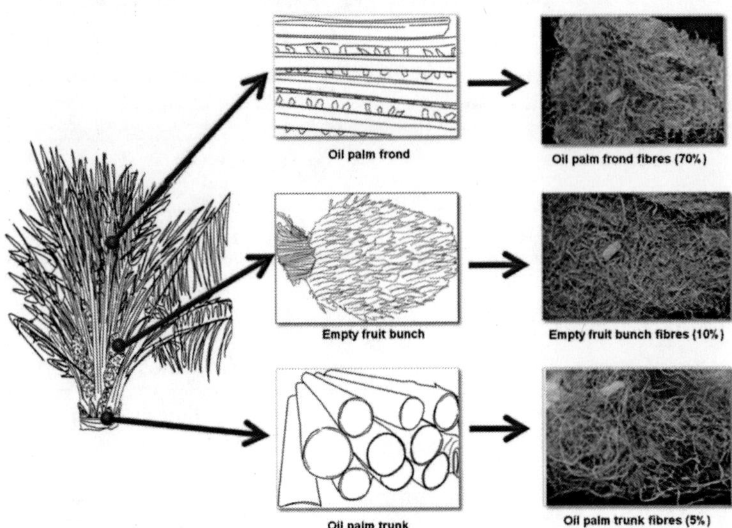


**Figure 3.** Oil palm biomass and oil palm biomass fibres from oil palm tree

Oil palm tree discarded for replantation after 25~30 years of oil production. Related to the large production of main products from oil palm in Malaysia, there is abundance of oil palm trunk. A large quantity of cellulosic raw material generated in the form of felled trunks during replanting can be utilized. Oil palm trunk obtained from oil palm tree and it consists of vascular bundles and parenchayma. Oil palm trunk amounted to approx 3 million tonnes/ year. Up to now, there is no economical value of oil palm trunk from the structural point of view and ultimately it becomes a hazardous material to farmers. To increase the added value of these residues, several investigations have been carried out to produce hybrid plywood, MDF, polymer composites, particle boards, paper, pulp, furniture, bio fuels etc. from oil palm biomass. Many studies have been carried out on the utilization of the oil palm EFB fibres such as in particleboard, pulp, medium density fibreboard, and composites (Rozman et al. 2005). For non-structural applications, works are continuing to look into the possibilities of using OPT as furniture and particleboard raw material (Chew 1985). At present, most of the oil palm biomass are disposed off at the oil palm plantation or burned at the mills to produce oil palm ash. Thus, finding useful utilization of the oil palm biomass in fabrication of natural fibre based composites/hybrid composites will surely alleviate environmental problems related to the disposal of oil palm wastes. Application of oil palm biomass fibre in different sector is shown in Table 1.

**Table 1.** Application of oil palm fibers

| Oil palm biomass | **Products** | **References** |
|---|---|---|
| **Oil palm EFB fibres** | Plywood | Abdul Khalil et al. 2010c |
| | MDF | Abdul Khalil et al. 2008; Abdul et al. 2010b |
| | Polymer biocomposite | Chai et al. 2009 |
| | Hybrid composite | Jawaid et al. 2010 |
| | Particle boards | Zaidon et al. 2007 |
| | Biofuel | Shuit et al. 2009 |

| | | |
|---|---|---|
| **Oil palm frond fibres** | Pulp, Paper<br>Nutrient recycling<br>Fibreboard<br>Biodegradable film<br>Animal feed<br>Downdraft gasifier | Wan Rosli et al. 2004; Abu Hassan 1994<br>Abu Hassan 1994<br>Abu Hassan 1994<br>Noor Haliza 2006<br>Hassim et al. 2010; Dahlan 2000<br>Sulaiman 2011 |
| **Oil palm trunk fibres** | Lignin<br>Plywood, Furniture | Xiao et al. 2001<br>Abdul Khalil et al. 2010c; Abdul Khalil et al. 2010a; Chew 1985 |

## Chemical Composition of Oil Palm Fibers

It is well know that chemical constituents of oil palm biomass significantly vary due to their diverse origins and types (Chew and Bhatia 2008). Oil palm biomass is lignocellulose residues composed of cellulose, hemicellulose, lignin and ash (Raveendran et al.1995). Table 2 shows the chemical composition of different oil palm biomass. All the oil palm biomass are rich in lignin, cellulose and hemicellulose (Meier and Faix 1999; Demirbaş 2000). Oil palm EFB fibers are lignocellulose fibres where the cellulose and hemicellulose are reinforced in a lignin matrix similar to that of other natural fibres. These oil palm empty fruit bunch consist of high cellulose content and is a potential natural fibre resources, but its applications account for a small percentage of the total biomass productions. High cellulose content and high toughness value of oil palm EFB fibres make it suitable for application in polymer composites (Sreekala et al., 2004; John et al., 2008). The cell wall of OPF also consists of cellulose, hemicellulose and lignin. In addition to these main components, ash, glucose and xylose are also present in cell wall of oil palm fibres. It was revealed that oil palm fibre from oil palm frond contain highest composition of hemicellulose compared to coir, pineapple, banana, and even soft and hardwood fibres (Abdul Khalil et al. 2006). Cellulose, hemicellulose, and lignin that form major constituents of the oil palm EFB fibre might differ depending on plant age, environment, soil condition, weather effect, and testing methods used (Table 3). Researchers reported that chemical composition of

EFB and OPF are quite comparable with coir but lower in cellulose content as compared to jute and flax fibres (Khalil et al. 2000). Oil palm trunk fibre is strong and high content of lignin (23%) as lignified cellulose fibres retain their strength better than delignified fibres. The chemical compositions of a lignocellulosic fibre vary according to the species, growing conditions, method of fibre preparations and many other factors (Bledzki and Gassan 1999).

**Table 2.** Chemical composition of oil palm biomass

| Composition | **Oil Palm biomass chemical composition (wt%)** | | |
|---|---|---|---|
| | **Oil Palm EFB** | **Oil palm Frond** | **Oil Palm Trunk** |
| Cellulose | 43-65 | 40-50 | 29-37 |
| Hemicellulose | 17-33 | 34-38 | 12-17 |
| Holocellulose | 68-86 | 80-83 | 42-45 |
| Lignin | 13-37 | 20-21 | 18-23 |
| Xylose | 29-33 | 26-29 | 15-18 |
| Glucose | 60-66 | 62-67 | 30-32 |
| Ash | 1-6 | 2-3 | 2-3 |

**Sources:** Law et al., 2007; Abdul Khalil 2006;Punsuvon 2005); Shinoj et al. 2011; Chew and Bhatia 2008.Mohamad and Abdul Halim 1985; Abdul Khalil 2004);Law and Jiang 2001); Sreekala et al. 2001

**Table 3.** Chemical composition of oil palm EFB fibres from different researchers

| Hemicellulose (%) | Cellulose (%) | Lignin (%) | Location | References |
|---|---|---|---|---|
| 22 | 48 | 25 | Malaysia | Hill and Abdul Khalil 2000 |
| 30 | 36 | 22 | Indonesia | Minowa et al.1998 |

| | | | | |
|---|---|---|---|---|
| 30 | 50 | 18 | Malaysia | Khalil et al. 2008 |
| 28 | 65 | 19 | India | Sreekala et al.1997; Law et al. 2007 |

## Physical properties of oil palm fibers

Table 4 provides data on physical properties of different oil palm biomass fibers. Fiber strength is a crucial factor to choose fibre that is specific for certain usage. Fibre length is an vital factor in determining bonding and stress distribution (Khalil et al. 2008). Oil palm EFB fiber length values between hardwood and softwood fibre length (Hassan et al. 2010). Researchers also reported that aspect ratio (l/d) of fibre has a significant effect on the properties of final composite materials. Researchers also reported that OPF fibres are shorter and thicker as compared with EFB and OPT fibres. Fibre with thicker cell wall resists collapse and do not contribute to interfere bonding to the same extent (Reddy and Yang 2005). The micro fibril angle, cell dimensions, and the chemical composition of fibres are the essential variables that determine the over all properties of the fibres (John and Thomas 2008). OPT fibres show high density which also indicates that fibre is strong. Researchers reported that fibers with higher lignin content, lower l/d ratio and higher microfibrillar angle show lower strength and modulus but have higher extensibility (Reddy and Yang 2005). The shape and size of cell lumen depends on the cell wall thickness, source of the fibres and it affect the bulk density of fibres (Reddy and Yang 2005). The physical properties such as length, diameter, lumen width, density and microfibril angle of oil palm fibres have revealed momentous changes in the physical and mechanical properties of the composite materials (Hassan et al. 2010; Shinoj et al. 2011; Jawaid and Abdul Khalil 2011a). Owing to their low specific gravity, which is about 1.25–1.50 g/cm$^3$ as compared to glass fibers which is about 2.6 g/cm$^3$, the lignocellulosic fibers are able to provide a high strength-to-weight ratio in plastic materials (Abu Bakar et al. 2005).

**Table 4.** Physical properties of oil palm biomass fibers

| Fibre | Fibre length (mm) | **Fibre Dia ((m)** | Lumen width (m) | Density (g/cm³) | **Fibril Angle (⁰)** | Ref |
|---|---|---|---|---|---|---|
| Oil Palm EFB | 0.89-142 | 8 –300 | 8 | 0.7-1.55 | 46 | Mohamad and Abdul Halim 1985; Law and Jiang 2001; Bismarck 2005; Amar 2005),Zulkifli et al. 2009),Khalil et al. 2008,Hassan et al. 2010 |
| Oil Palm Frond | 0.59 -1.59 | 11-19.7 | 8.20-11.66 | 0.6-1.2 | 40 | Mohamad and Abdul Halim 1985; Law and Jiang 2001; Amar 2005, Khalil et al. 2008, Law and Jiang 2001 |
| Oil Palm Trunk | 0.60 -1.22 | 29.6 –35.3 | 17.60 | 0.5-1.1 | 42 | Mohamad and Abdul Halim 1985; Khoo 1985; Amar 2005;Khalil et al. 2008;Ahmad et al.2010 |

## Mechanical Properties of Oil Palm Biomass Fibers

Table 5 gives the data for mechanical properties of oil palm biomass fibres. Mechanical properties such as tensile strength and modulus related to the composition and internal structure of the fibers. It reported that generally the tensile strength and young's modulus of plant fibre increases with increasing cellulose content of the fibres (Aji et al. 2009). Oil palm trunk fibre found to be suitable as reinforcement because it possesses high tensile strength (300-600

MPa) which is considered high when compared with other natural fibre. The properties of cellulosic fibers are strongly influenced by chemical composition, fibre structure, microfibril angle, cell dimensions and defects, it differs from different parts of a plant as well as from different plants (Dufresne 2008). The thicker walled fibre tend to produce an open and bulky sheet with low burst/ tensile strength and high tearing resistant (Mishra et al. 2004). The mechanical properties of natural fibers also depend on their own cellulose and its crystalline organization, which can determine the mechanical properties (Bledzki and Gassan 1999). Oil palm fibres are hard and tough and found to be a potential reinforcement in polymer composites (Jawaid and Abdul Khalil 2011 a).

**Table 5.** Mechanical properties of oil palm fibre

| Fibres | Tensile strength (MPa) | Young's modulus (GPa) | **Elongation at break (%)** | **Ref** |
|---|---|---|---|---|
| Oil Palm EFB | 50-400 | 0.57-9 | 2.5-18 | (Sreekala et al. 2004; Bismarck 2005; Kalam et al. 2005; Bakar et al. 2006) |
| Oil Palm Frond | 20-200 | 2-8 | 3-16 | Lab sources |
| Oil Palm Trunk | 300-600 | 8-45 | 5-25 | Ahmad et al. 2010; Killman and Hong 1989; Lim and Khoo 1986 |

## Oil Palm Fiber based hybrid composites

The oil palm fibers have been focus of study in Malaysia and around the world but still oil palm fibers have not found a solid economic value. Oil palm fibres are abundant in nature and due to its low density, non-abrasiveness and biodegradability can be used in hybrid composites. In order to take full advantage of the oil palm fibres, it can be combined with other high strength fibres in the same matrix to produce hybrid composites, and thereby an economically viable composite can be obtained. Researchers reported that fibre content, fibre length, orientation, extent of intermingling of fibres, fibre/ matrix interface and arrangement of both the fibres mainly affect the over all properties of a hybrid composite (Munikenche Gowda et al. 1999; Sreekala et al. 2002). Hybrid effect is defined as a positive or negative deviation of a certain mechanical property from the rule of mixture behaviour (Kickelbick 2007). Researchers explained that the rule of mixture defines a composite property as weighted average of the properties of its constituents. For example, if two fibres of different properties such as oil palm fibres and jute fibres are incorporated in the polymer, the resulting hybrid composite would most probably exhibit properties which are some sort of an average between those individual fibre components. Reported work on oil palm fibre based hybrid composites are shown inTable 6.

## PHYSICAL AND MECHANICAL PROPERTIES OF OIL PALM FIBRE BASED HYBRID COMPOSITES

The physical properties are used to calculate the product weight, which relates to manufacturing costs, injection molding machine melt capacity (machine size), and the product dimensional control (mold shrinkage). The mechanical properties of natural fibre composites depends on fibre properties, surface character of material, types of matrix, fibre-matrix bonding, volume fraction of fibre and alignment of fibers etc. Mechanical properties of natural fibre composite also depend on composite processing method and effectiveness of the coupling between the fibre and matrix phases. Most of the studies on natural fibre hybrid composite involve study of mechanical properties as a function of fibre length, fibre loading, extent of intermingling of fibres, fibre to matrix bonding and arrangement of both the fibres, effect of various chemical treatments of fibers, and use of coupling agents.

**TABLE 6.** Reported works on Oil Palm Fibre based Polymer Hybrid Composites

| Hybrid | Matrix | References |
|---|---|---|
| Oil Palm/Glass | Polyester<br><br>Polypropylene(PP)<br>Phenol<br>Formaldehyde(PF)<br>Epoxy<br>Vinyl Ester<br>Natural rubber | Kumar et al. 1997;Agrawal et al. 2000;Abdul Khalil et al. 2007; Karina et al. 2008; Wong et al. 2010)<br>Rozman et al.2001a,b<br>Sreekala et al. 2002a,b; Sreekala et al. 2004; Sreekala et al. 2005)<br>Hariharan et al. 2004; Abu Bakar et al. 2005; Mridha et al. 2007<br>Abdul Khalil et al. 2009<br>Anuar et al. 2006 |
| Oil palm/Sisal | Natural rubber | Jacob et al. 2004a,b; Jacob et al., 2005;Jacob et al. 2006a,b,c,d;Jacob et al. 2007; John et al. 2008 |
| Oil Palm/Jute | Epoxy | Jawaid et al., 2010;2011a-g; Khalil et al., 2011 |
| Oil Palm/ Kalonite | Polyurethane | Anuar and Badri, 2007 |

## OIL PALM/GLASS FIBRES REINFORCED HYBRID COMPOSITES

Hybridization of oil palm fibres with glass fibres carried out by several researchers in Malaysia, India and Indonesia. It's clear from the literature review that first time oil palm/glass hybrid composites fabricated and researchers developed fire resistant sheet moulding hybrid composites by measuring fire retardancy through limiting oxygen index (Kumar et al. 1997). After initiative from Kumar et al. (1997), another researchers carried out non-isothermal crystallization kinetics of oil palm/glass hybrid composites and analyzed it in the light of Ozawa's Theory (Agrawal et al. 2000). Preliminary study on mechanical properties of oil palm/glass fibres as reinforcing agents in polypropylene (PP) matrix was carried out (Rozman et al. 2001a,b). Results indicated that the incorporation of oil palm and glass fibres into PP matrix resulted in the reduction of tensile and flexural strength (Rozman et al. 2001a). They also studies the effect of coupling agents on tensile and flexural properties of hybrid composites and concluded that only slight improvements in some cases were shown for those composites treated with polymethylenepolyphenyl isocyanate. In another interesting research, they reported that effect of extraction of the oil palm EFB fibres show significant improvement in flexural and tensile strength and toughness, with a slight increase in the flexural and tensile modulus and elongation at break of oil palm/ glass hybrid composites (Rozman et al. 2001b).

Researchers reported extensive study on effect of glass fibre loading on tensile, flexural and impact response of oil palm EFB fibre/phenol formaldehyde(PF) composite (Sreekala et al. 2002a). The over all mechanical performance of the hybrid composite was improved except impact strength and density of the hybrid composite decreases as volume fraction of oil palm EFB fibre increases while hardness of the hybrid composite also showed a slight decrease on an increased volume fraction of oil palm EFB fibre in hybrid composites. Similar study carried out on the influence of glass fibre loading, relative volume fraction of fibres in hybrid composites, and surface modification of fibres on water sorption kinetics in oil palm/ glass fibres reinforced PF hybrid composites (Sreekala et al. 2002b). Results out put indicated that Hybridization of oil palm fibre with glass fibre considerably decreased the water sorption of oil palm composite. Accelerated weathering studies of untreated and treated oil palm/glass reinforced PF composites conducted and observed biodegradation and irradiation effects in light of variations in tensile and impact properties (Sreekala et al. 2004). Results indicated that mechanical performance of hybrid composites decreased with thermal and radiation ageing. Researchers studied physical and mechanical properties of the oil palm/glass fibres reinforced epoxy resin bi-layer and tri-layer hybrid composites (Hariharan et al. 2004; Abu Bakar et al. 2005). They observed that hybridization of oil palm with glass fibres increased impact strength of the hybrid composites but tensile strength and young's modulus show negative hybrid effect while positive hybrid effect was observed for the elongation at break of the hybrid composites.

In 2006, one researchers studied tensile and impact properties of thermoplastic natural rubber hybrid composite with short glass fibre and oil palm EFB fibres for the first time (Anuar et al. 2006). Researchers focused on the effect of treatment of glass and oil palm fibres with coupling agent such as silane, and maleic anhydride grafted polypropylene (MAgPP). Obtained results show that hybrid composites containing 10% EFB and 10% glass fibre gave an optimum tensile and impact strength for treated and untreated hybrid composites and tensile properties increased with addition of coupling agent. In another study on physical (density and water absorption), and mechanical (tensile, flexural and impact) properties of oil palm EFB/glass fibres reinforced polyester hybrid composites investigated with increasing loading of both oil palm EFB and glass fibres (Abdul Khalil et al. 2007). Hybridization of glass fibres with oil palm EFB fibres improved mechanical and physical properties of hybrid polyester composite as compared to oil palm EFB/Polyester composites due to the high strength and modulus value of glass fibre than the inferior EFB fibre. In 2007, first type researchers hybridized oil palm wood flour particle (Size < 250m) with woven glass fibre reinforced epoxy composite and fabricate hybrid composites by hand lay-up method (Mridha et al. 2007). They reported that impact strength reduced with increasing the filler content and also several damages found in specimens at higher filler content resulting higher energy absorption during impact. Similar work reported by Indonesian researchers who studied physical and mechanical properties of oil palm/glass fibre reinforced polyester hybrid composites related to EFB fibre specimen length and fibre loading (Karina et al. 2008). Results show that oil palm fibre length did not show any significance effect on the flexural strength and density of hybrid composites but shorter EFB fibre show low dimensional stability as compared to longer

Until now only one research paper reported on mechanical and physical properties of the oil palm EFB/glass fibre reinforced vinyl ester hybrid composites at a different layer arrangement (Abdul Khalil et al. 2009). Mechanical and physical properties of hybrid composites were found higher than that of mechanical and chemical treated oil palm fibres reinforced composites. It also clear from the results that layering pattern of oil palm EFB and glass fibres within hybrid composites affect dimensional stability and decrease by the incorporation of glass fibre as compared to mechanical and chemical treated composites. Recently conducted a study on impact behaviour of E-glass/oil palm fibres reinforced polyester hybrid composites showed that impact strength improved with increasing number of glass fibre layer and increment in fibre length (Wong et al. 2010).

## OIL PALM/SISAL FIBRES REINFORCED HYBRID COMPOSITES

In 2004, Ist time any researchers try to fabricate oil palm based hybrid biocomposites by hybridization of oil palm fibres with other natural fibres by the unique combination of sisal and oil palm fibres reinforced rubber composites (Jacob et al. 2004a,b). Researchers studied the effect of fibre loading, fibre ratio, and treatment of fibres on mechanical properties of sisal/oil palm fibre reinforced hybrid composites (Jacob et al. 2004a). Results indicated that increasing the concentration of fibres reduced tensile and tear strength, but enhanced tensile modulus of the hybrid composites. They also reported that 21g sisal and 9 g oil palm based hybrid composite show maximum tensile strength and concluded that tensile strength of hybrid composites depend on weight of sisal fibres rather than oil palm fibres due to high tensile strength of sisal fibres. It also seen that treatments of both sisal and oil palm fibres causes better fibre/matrix interfacial adhesion and resulted in enhanced mechanical properties. Similar studies carried out on the influence of fibre length on the mechanical properties of untreated sisal/oil palm fibre based hybrid composites and reported that increase in fibre length decreases the mechanical properties of hybrid composites due to fibre entanglements (Jacob et al. 2004b).

Researchers also studied water sorption characteristics of the oil palm/sisal hybrid composites with reference to fibre loading, chemical modification and influence of temperature (Jacob et al. 2005). Mercerization and silanation treatment of sisal and oil palm fibres decreased the water uptake in the hybrid composites and moisture uptake found to be dependent on the properties of the biofibres. In an interesting study, researchers reported durability and ageing characteristics of oil palm/sisal hybrid composites (Jacob et al. 2007).

Researchers again fabricated oil palm /sisal hybrid composites by using surface treated oil palm and sisal fibre with varying concentration of NaOH solution and different coupling agents (John et al. 2008). They compared fibre reinforcing efficiency of the chemically treated biocomposites with that of untreated composites. Results demonstrated that chemical treatment of sisal and oil palm fibres enhanced mechanical properties of hybrid composites and 4% NaOH show superior tensile properties due to better fibre/matrix interaction. Hybrid composites prepared from fluorosilane treated fibres exhibited better mechanical properties as compared to other silane treated hybrid composites. They also investigated anisotropic swelling studies of hybrid composites and revealed that fluorosilane treated fibres based hybrid composites has the highest degree of fibre alignment.

## OIL PALM/JUTE FIBRE REINFORCED HYBRID COMPOSITES

Recently we carried out an extensive study on mechanical and physical properties of oil palm/jute fibres reinforced epoxy hybrid composites in our laboratory. Research output show that dimensional stability, density, void content, and chemical resistance of oil palm EFB fibres reinforced epoxy composite improved with hybridization of oil palm EFB fibres with non woven/woven jute fibres (Jawaid et al. 2011d,f,g; Abdul Khalil et al. 2011). Hybridization of oil palm EFB composites with jute fibres improved physical properties of oil palm/jute hybrid composites due to packed and hybrid arrangement of fibre, less hydrophilic nature of jute fibres, better jute/epoxy interaction. It also observed that woven hybrid composites display poor dimensional stability, density and chemical resistant due to high void content of woven jute fabrics as compared to non woven fibres.

Mechanical performance of oil palm EFB fibres with non woven/ woven jute fibres reinforced hybrid composites were also carried out (Jawaid et al. 2010; 2011b,c,e). Results indicated that oil palm/ non woven jute fibre based hybrid composites has higher tensile and flexural properties due to high tensile strength of jute fibres and better jute/epoxy matrix interaction. Hybrid composite show less impact strength as compared to oil palm EFB composite because oil palm EFB fibre has high fracture toughness as compared to jute fibres (Jawaid et al. 2010; 2011c,g). It observed that mechanical properties of woven hybrid composites significant improved work as compared to non woven hybrid composites due to woven fibres are tightly bonded, low deformation at break, stretching nature of fabrics or fabrics mats have higher fibre count than chopped strand (non woven) mat (Jawaid et al. 2011b,e).

## OIL PALM/KAOLINITE HYBRID COMPOSITES

Mechanical and water sorption properties of oil palm empty fruit bunch fibres and kaolinite reinforced polyurethane hybrid bio-composites were studied (Amin and Badri 2007). Hybrid bio-composites showed maximum flexural and impact strengths were at 15% kaolinite loading. It observed that oil palm fibre hybridization with kaolinite improved stiffness, strength, and display better water resistance to extent. Mechanical properties of oil palm/Kaolinite hybrid enhanced due to interaction between kaolinite with PU matrix and EFB fibres.

## ELECTRICAL PROPERTIES OF OIL PALM FIBRE BASED HYBRID COMPOSITES

In the field of natural fibre reinforced polymer composites, extremely limited research has been reported on the electrical properties. In 1981, Ist time researchers reported electrical resistivity and dielectric constant of some natural fibre (Coir, banana, sisal pineapple leaf, and palmyra fibres) and results indicated that natural fibre has high electric resistivity and dielectric strength (Kulkarni et al. 1981). They also suggested that natural fibre can be used as a replacement for wood in insulating applications. Further study reported by another researcher on electrical resistivity of various part of the coconut tree and confirm that it can used as good insulators (Satyanarayana et al. 1982). Recently electrical properties of oil palm fibres reported and concluded that dielectric constant value of oil palm fibre is in agreement with the values reported earlier (Chand 1992) for other natural fibres (Shinoj et al. 2010). Results obtained by researchers also clear that fibre size reduction caused an increase in dielectric constant, while alkali treatment on fibres caused a decrease in dielectric constant. Several researchers worked on electrical properties of natural fibre reinforced polymer composites (Reid et al.1986; Paul and Thomas 1997; Datta et al.1984; Jayamol et al.1998; Chand and Jain 2005). Researchers studied electrical properties of untreated and treated oil-palm fibre reinforced polymer composites by using transient plane source technique at room temperature. Results indicated that all the silane and alkali treated fibres shows enhancement in thermal conductivity and diffusivity of the composites as compared to acetylated composite (Agrawal et al. 2000). In this work, the dielectric properties of oil palm/natural rubber composite materials of untreated and treated oil palm fibres were characterized. It noticeable that oil palm fibre reinforced in rubber matrix affects the dielectric properties of composites and treatment of fibres increase the loss factor. They also reported that low fibre content (20%) show less significant effect on the dielectric properties of composites (Marzinotto et al. 2007). Recently, another researchers investigate the influence of the fibre orientation in the dielectric properties of the oil palm tree fibre reinforced polyester composite by using a dielectric spectroscopy (Ben Amor et al. 2010). Results demonstrated that orientation of the fibre can strongly influence the dielectrical properties and interfacial polarization processes in composites and this technique can be used to probe the composite interphase and investigate the effect of fibre orientation on the evolution of composite interfacial properties.

In an attractive research work carried out on thermal conductivity and diffusion of banana/glass fibres reinforced rubber hybrid composites in relation to fibre loading, fibre ratio, frequency, chemical modification of fibres and the presence of a bonding agent (Agarwal et al. 2003). Electrical strength and volume resistivity of jute/wheat and jute/jamun fibres reinforced bispheniol-c-formaldehyde has been evaluated (Mehta and Parsania 2006). Researchers reported that there is no much difference in dielectric breakdown strength between hybrid composites but volume resistivity of hybrid composites increased by 197-437% as compared to jute fibres reinforced bispheniol-c-formaldehyde composite. Researchers investigated dielectric behaviours of glass and jute fibres reinforced polyester hybrid composites and later on electrical properties of banana/

glass hybrid fibre reinforced composites with varying hybrid ratios and layering patterns were analyzed (Fraga et al. 2006; Joseph and Thomas 2008). In 2006, Ist time dielectric characteristics and volume resistivity of sisal-oil palm hybrid biocomposites investigated (Jacob et al. 2006). Results obtained show that dielectric constant increases with fibre loading at all frequencies, and volume resistivity decreases with frequency, and fibre loading, this implies that the conductivity increases upon addition of lignocellulosic fibres. Researchers also reported that chemical modification of fibres resulted in a decrease in dielectric constant and increase in volume resistivity due to increase in hydrophobicity of fibres. It also observed that addition of a two-component dry bonding agent consisting of hexamethylene tetramine and resorcinol, used for the improvement of interfacial adhesion between the matrix and fibres reduced the dielectric constant of the hybrid composites. They also reported that dissipation factor was seen to increase with fibre loading which indicates that the electrical charges can be retained over a longer period of time. Researchers studied dielectric constant and volume resistivity values of sisal/ coir hybrid composites and reported similar results to oil palm/sisal hybrid composites (Haseena et al. 2007).

## THERMAL AND DYNAMIC MECHANICAL PROPERTIES OF OIL PALM FIBRE BASED HYBRID COMPOSITES

Thermal stability of natural fibres and natural fibre based composites can be analyzed by two techniques viz., thermogravimetric analysis (TGA) and differential scanning calorimeter (DSC). Weight and energy loss with heating is common phenomena for natural fibre reinforced polymer composites due to degradation and loss of residual solvents and monomers. Weight loss on heating is studied by TGA and measurement of relative changes in temperature and heat or energy either under isothermal or adiabatic conditions studied by DSC. Dynamic mechanical analysis (DMA) is a technique for measuring the modulus and damping factor of a sample. The modulus is a measure of how stiff or flimsy sample is, and amount of damping a material can provide is related to energy it can absorb (Duncan 2008). DMA is generally used for thermoplastics, thermosets, composites and biomaterials. DMA is one of the most powerful tools to study the behaviour of polymer composite materials and it allows for a quick and easy measurement of material properties (Swaminathan and Shivakumar 2009). Previously thermal properties of hybrid composites such as banana/glass, sisal/glass, bamboo/glass, hemp/glass, etc. were studied by researchers and observed that addition of glass fibre improved thermal properties of natural fibre based composites. Researchers reported that hybridization of banana fibre reinforced polymer composites with glass fibres enhanced melting point, crystallization temperature and onset thermal degradation temperature of maleic anhydride grafted polypropylene (MAPP) treated banana/glass hybrid composites (Samal et al. 2009a). It is due to SiO groups in glass fibre which interlinks with anhydride group of MAPP, providing synergism between glass and banana fibres. Similar study on thermal stability of banana/glass hybrid composites carried out by TGA and DSC and revealed that MAPP treated banana and glass fibres enhanced thermal stability of polypropylene (Nayak et al. 2010a). Dynamic mechanical properties of banana/glass hybrid composites have been analyzed to investigate the interfacial properties (Samal et al. 2009a;Nayak et al. 2010a). Results show that the storage modulus and loss modulus of MAPP treated hybrid composites improved over the whole temperature range, indicating better adhesion between fibre/matrix. Dynamic mechanical properties of banana/glass hybrid composites also studied over a range of temperature and three different frequencies (Pothan et al. 2010). Storage modulus values of hybrid composite decreases above the glass transition temperature (Tg) where glass is the core material.

Researchers studied thermal properties of sisal/glass hybrid composites and confirm that addition of glass fibre improved thermal properties of sisal/PP composites (Jarukumjorn and Suppakarn 2009;Nayak and Mohanty 2010b). In an interesting study on thermal properties such as crystallization, melting behaviour and thermal stability of bamboo/glass hybrid composites and obtained results indicate an increase in thermal stability of bamboo composites due to the higher thermal stability of glass fibre than bamboo fibre (Samal et al. 2009b; Nayak and Mohanty 2010b; Lee and Wang 2006). Hybridization of hemp fibre composites with glass fibres shifts the temperature of degradation to a higher value indicating an increased thermal stability of the hybrid composites (Panthapulakkal and Sain 2007). Effect of glass fibre loading and MAPP treatment on dynamic mechanical properties of sisal/glass hybrid composites carried out and reported improvement in storage modulus and loss modulus of hybrid composites with hybridization of sisal/pp composite with glass fibres and treatment with MAPP (Ornaghi Jr et al. 2010; Nayak and Mohanty 2010b). Dynamic mechanical properties of bamboo/glass fibre hybrid composites were also studied and observed results indicate an increase in storage modulus and decrease in damping properties due to hybridization and treatment of fibres with MAPP as compared to untreated composites and pure matrix (Nayak et al. 2009, 2010c; Samal et al. 2009b). Until now, there is no any researcher work carried out on thermal stability of banana/sisal and pineapple leaf/glass reinforced hybrid composites but few work reported on dynamic mechanical properties of these hybrid composites. They concluded that storage modulus and damping factor of banana/sisal and pineapple leaf/glass hybrid composites also enhanced due to hybridization with natural and synthetic fibres (Idicula et al. 2005a,b; Uma Devi et al. 2010).

Literature review indicated that particularly limiting work reported on thermal and dynamic mechanical properties of oil palm based hybrid composites. In 2006, Ist time any researchers studied thermal properties of sisal/oil palm fibres reinforced natural rubber hybrid composites and observed that sisal fibre loading and chemical modification of fibres enhanced thermal stability of the oil palm composites (Jacob et al. 2006b). In case of oil palm/sisal hybrid composite, the peak temperatures have decreased to 356.3$^0$C, and a new peak has come at 489.9$^0$C due to hemicellulose and alpha-cellulose degradation and addition of fibers results in an increase of thermal stability as indicated by the higher peak temperatures. Chemical modification of fibres results in a further increase of thermal stability as evident from the peak temperatures of 524$^0$C and 518.3$^0$C of 4%NaOH and aminosilane treated composites.

They also carried out an extensive study on dynamic mechanical properties of oil palm/sisal fibre reinforced natural rubber hybrid composites as a function of temperature (Jacob et al. 2006c). The storage modulus of hybrid composites increased with sisal fibre loading while the damping factor found to decrease due to the increased stiffness imparted by oil palm and sisal fibres. Treatment of fibres with NaOH also enhanced the storage modulus value of hybrid composites while chemically treated composites show decreased in damping factor value as compared to untreated composites. Similar study on dynamic mechanical properties of oil palm/sisal hybrid composites with reference to the role of coupling agents were carried out (Jacob et al. 2006d). It observed that storage and loss modulus of treated hybrid composites increased while damping properties decreased due to better interfacial interface between fibre and matrix. The dynamic mechanical properties of oil palm fibre/glass fibre reinforced phenol formaldehyde hybrid composites as a function of fibre content, and hybrid fibre ratio was investigated (Sreekala et al. 2005). The glass transition temperature of the hybrid composites found to be lower than that of the unhybridized composites. Storage modulus of the hybrid composites was also found to be lower than that of unhybridized oil palm fibre/PF composite. Loss modulus of hybrid composites increased with increase in glass fibres, and gradual decrease in loss modulus is observed with the increase in frequency. It also noted that the activation energy decreases with incorporation of glass fibres in hybrid composites.

In our laboratory, we carried out research work on thermal and dynamic mechanical properties of oil palm/jute fibre reinforced epoxy hybrid composites (Jawaid and Abdul Khalil 2011 e; Jawaid et al. 2012). The thermal behaviour of the oil palm/jute based hybrid composites were determined (Figure 4), and it noticed that initial degradation temperature of hybrid composites shifted to a higher temperature well over 268-271°C as compared with the EFB composite (260°C), which indicate higher thermal stabilities of the hybrid composites (Jawaid and Abdul Khalil 2011 e). Final degradation temperature of hybrid composites also shifted between 441 to 443°C due to complex reaction. Hybridization of oil palm EFB fibres with jute fibres, the final decomposition temperature and ash content of the hybrid composite shifted slightly towards higher temperature as a result of the high thermal stability of jute fibre which acts as barriers to prevent the degradation of oil palm EFB fibres. Similar study carried out on thermal stability of oil palm/woven jute fibre reinforced epoxy hybrid composites and results indicated that woven hybrid composites initial and final decomposition temperature higher than oil palm/jute hybrid composites (Jawaid et al. 2012). It clear from Figure 4that oil palm EFB/woven jute hybrid composites are more thermal stable as compared to oil palm EFB composite which is possibly due to the higher thermal stability of woven jute fibre than oil palm EFB fibre.

Dynamic mechanical properties of oil palm/jute and woven hybrid composites also carried out in our laboratory (Jawaid and Abdul Khalil 2011 e; Jawaid et al. 2012). Figure 5 shows storage modulus values of hybrid composites at a frequency of 1 Hz. On investigating the variation of storage modulus with temperature, storage modulus was found to be decreased with the increase in temperature in all cases at low temperature because fibre do not contribute much to imparting stiffness to the material. As temperature increases, the components become more mobile and lose their mobility and lose their close packing arrangement. The high stiffness of hybrid composites is in agreement with their tensile property reported in our previous research (Jawaid et al. 2011g). In particular the composite stiffness is substantially increased with woven jute and EFB fibre incorporation, and it causes a drop in the modulus above $T_g$ and comparatively less than epoxy matrix (Jawaid et al. 2012). The stiffness of the oil palm EFB composite increases with hybridization with woven jute fibres, resulting in high storage modulus (Figure 5).Moreover, the addition of woven jute fibres allows greater stress transfer at the interface and ultimately increases the storage modulus.

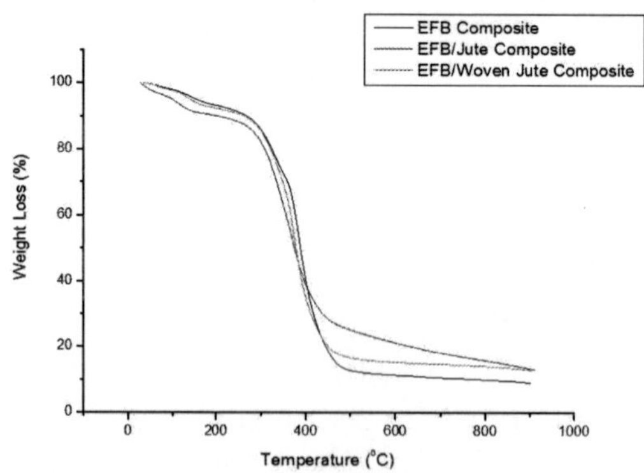


**Figure 4.** Thermogravimetric analysis curve of Oil Palm EFB, oil palm EFB/ jute, and Oil Palm EFB/Woven Jute Hybrid Composites.

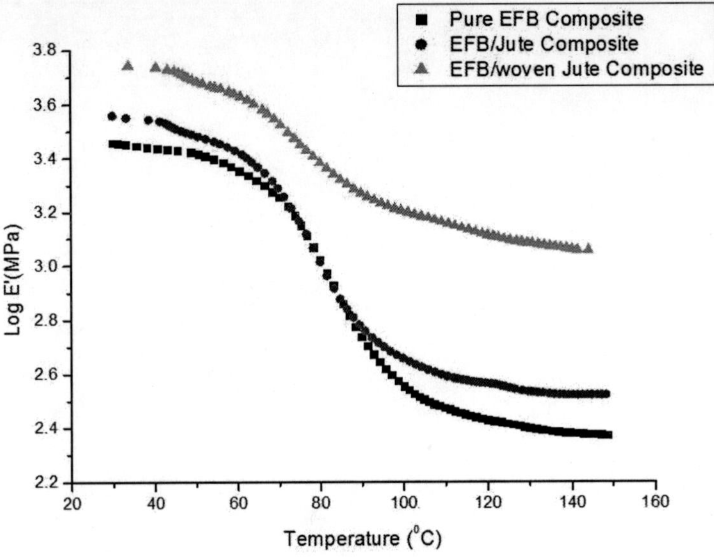


**Figure 5.** Storage modulus of oil palm EFB, oil palm EFB/jute, and oil palm EFB/woven jute hybrid compossites

Trends of change in the damping factor of the hybrid composites with temperature are shown inFigure 6. Damping factor of oil palm/non woven jute hybrid composite indicate that incorporation of the small amount of jute fibre to oil palm EFB/epoxy composite enhances the damping characteristics of the hybrid composites (Jawaid and Abdul khalil 2011e). With incorporation of oil palm EFB and woven jute fibres, the Tan δ peak was lowered as expected (Jawaid et al. 2012). Reinforcement of fabrics results in the formation of barriers that restrict the mobility of polymer chain, leading to lower flexibility, lower degree of molecular motion and hence lower damping characteristics (Jacob et al. 2006).

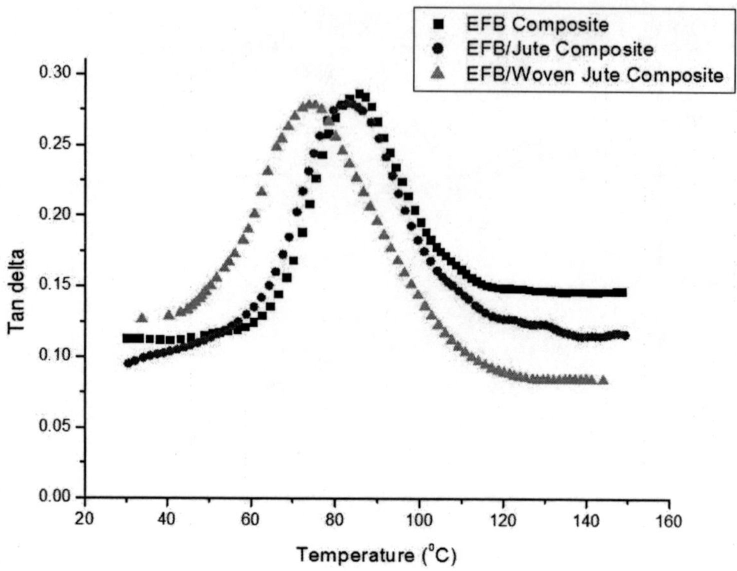


**Figure 6.** Damping factor of oil palm EFB, oil palm EFB/jute, and oil palm EFB/woven jute hybrid composites

## APPLICATION OF OIL PALM FIBRES BASED COMPOSITES

The use of bio-fibres for composite applications is being investigated throughout the world (Bledzi et al, 2006). Biocomposites have received increasing attention from both of the academic world and industries as building, automotive, packaging, and so on. Green composites or biocomposites based on natural fibres and resins are increasingly used for various applications as replacements for non-degradable materials. Green composites made entirely from renewable agricultural resources could offer a unique alternative for building, construction, furniture and automotive application (Gupta 2009). Biocomposites find applications in a number of fields viz., automotive industry and construction industry such as for panels, frames, ceilings, and partition boards (Abu Bakar et al. 2005), structural applications, aerospace, sports, recreation equipment, boats, office products, machinery, etc. (Sreekala et al. 2002a), and structural and non structural components such as window, door, siding, fencing, roofing, decking, etc. (Yatim et al. 2011). Bledzki et al. (2006) point out the importance of natural fibres and natural fibres composites in the automotive industry. Researchers already developed many automotive components (interior and exterior) from bio-fibre reinforced composite materials and designed Eco car which is built from bio fibre composite panels that incorporate biodegradable resins as the matrix material. (Bledzi et al, 2006). A modern automobile requires recyclable or biodegradable parts because of the increasing demand for a clean environment (Gupta 2009). Composites are also finding application in electrical and electronic industries.

Recently trends developed in the automotive industry to use natural and recycled materials to fulfil "The End of Life Vehicle (ELV) Directive", which came into force in October 2000, requires all member states to de-pollute all scrapped vehicles, avoid hazardous waste and reduce the amount of waste placed in landfill sites to a maximum of five per cent per car by 2015. Ford automotive stated to develop one million Lincoln and Mercury vehicles that feature recycled plastic bottles and even worn-out jeans in their seat cushions and backs by the end of the year, while French carmaker Citroën has turned to natural materials derived from hemp, bamboo and kenaf for the door linings of its 'premium' models. A team at Baylor University in the USA has made trunk liners, floorboards and car-door interior covers using fibres from the outer husks of coconuts, replacing the synthetic polyester fibres typically used. So, the main driving force to use natural fibres or recycled materials to develop cost effective, low weight and sustainable car to fulfil ELV directive and reduce hazardous waste.

Oil palm fibres can be used as fillers in thermoplastics and thermoset composites. These composites have wide applications in furniture and automobile components. Malaysian Palm Oil Board (MPOB) has developed a series of technologies on manufacturing thermo-formable plastic composites through compression or extrusion process for car components such as bumbers, trimmings, rear parcel shell, spare wheel cover and splash shield and also plastic pellets. Blending of oil palm fibres with polyols can produce cost effective lighter products such as packaging materials and roof insulators, which only require low compressive hardness. Oil palm fibre-filled automotive upholstery parts, dampening sheets for automotive industry using oil palm fibres, moulded particleboard, pulp and paper from oil palm fibres also developed. Researchers developed composite plastic fiber for automotive application by using oil palm biomass (Viva 2011). Progress in utilization of oil palm fibre finally reached to commercialization stage when PROTON (Malaysian National Carmaker) entered into an agreement with PORIM (Palm Oil Research Institute of Malaysia) to develop the thermoplastic and thermoset composites and used it in PROTON car (Panigrahi 2010). Sheffield University (UK) research team started to work with a producer of palm oil fibre and show interest in sustainable car production by replacing traditional materials and say that once the research will be in a later stage they will try to motive car manufacturers to use new materials and they hope that once the regulations become more stringent, this will force to do the change. In an interesting research, oil palm EFB fibre reinforced concrete roof slates produced, and preliminary results suggest that oil palm EFBs have a potential to be a material component of reinforced mortar roofing slates with the appropriate water cement ratio and concrete mix (Kaliwon 2010).

Researcher have developed, manufactured and assembled a small prototype car with all body panels made from jute fibre reinforced composite and hybrid composite. It also reported that door panels manufactured from flax/sisal hybrid epoxy composite shown remarkable weight reduction of about 20% (Schuh 2004). In 2000, Audi launched the A2 midrange car in which door trim panels were made of polyurethane reinforced with mixed flax/sisal mat. Researcher developed bamboo/glass fibre biocomposites for applications in wall panels, roofing panels, floors (Van Rijswijk et al. 2001). From literature view, we unable to find any application of oil palm fibres based hybrid composites which is required to explore the application in automobile, construction, packaging, etc sectors.

## CONCLUSIONS

Oil palm biomass is an agricultural by-product periodically left in the field after oil extraction and considers as hazardous material creating environmental problems. The oil palm biomass residue include the oil palm trunks, oil palm fronds, kernel shell, empty fruit bunch, presses fruit fibre, and palm oil mill effluent. Oil palm fronds accounts for 70%, EFB accounts for 10%, and OPT accounts for only about 5% of the total biomass produced. Oil palm biomass residues can be utilized as by-products, and it can also help to reduce environmental hazards. Oil palm biomass fibres consist of cellulose, hemicelluose, and lignin while oil palm empty fruit bunch consist of high cellulose content and is a potential natural fibre resource, but its applications account for a small percentage of the total biomass productions. The properties of oil palm fibres are strongly influenced by chemical composition, fibre structure, microfibril angle, cell dimensions and defects, it differs from different parts of a plant as well as from different areas. Oil palm biomass fibres offer excellent specific properties and have potential as outstanding reinforcing fillers in the matrix and can be used as an alternative material for bio-composites, hybrid composites, pulp, and paper industries. Natural fibres such as oil palm biomass fibres are not suitable for high performance applications due to its low strength, environmental sensitivity, and poor moisture resistance which results in degradation in strength and stiffness of natural fibre reinforced composites. Most of the drawbacks that have been identified can be overcome by effective hybridization of oil palm fibres fibre with glass fibre or natural fibres (jute/sisal).

The primary advantages of using oil palm fibres in hybrid composites are its low densities, non abrasiveness and biodegradability. Oil palm fibres with jute/sisal fibres appear to be promising materials because of the high tensile strength of jute and sisal fibres and the toughness of oil palm fibre. Now most of automobile manufacturer try to replace synthetic fibres with natural fibres, but it's not comparable in properties while fabricating hybrid composites by the combination of two natural fibres give them advantage to replace synthetic fibres. Hybridization of oil palm EFB fibres with glass/jute/sisal fibres enhanced physical and mechanical properties of oil palm EFB composites. It would appear that a wide variety of work carried out worldwide on physical and mechanical properties of oil palm/glass, oil palm /sisal, and oil palm/jute hybrid composites but still not fully explore electrical, thermal, and dynamic mechanical properties of oil palm based hybrid composites. Over all conclusions is that hybrid composites fabrication by using oil palm fibres will help in development of unique cost effective advanced composites possessing superior mechanical properties, dimensional stability, appropriate stiffness, damping behaviour and thermal stability.

Extensive research still required to do on oil palm based hybrid composite while exploring compatible of oil palm fibres with other natural fibres by compounded with several other polymers in the line of previous work. Microstructure of the interface between the oil palm fibre with other natural fibres, and matrix needs to be investigated and the interfacial properties should be studied with single fibre pull out, micro-bond test and single fibre fragmentation test. Complementary techniques such as X-ray photoelectron spectroscopy, time-of-flight secondary ion mass spectrometry may be studied to get more information about the chemical composition of the surface, its wetting behaviour, etc., that, coupled with the mechanical assessment of the interface, can shed more light on the structural characteristics of the interface. Future research on oil palm fibre based hybrid composites not only limited to its automotive applications but it also required to explore its application in aircraft components, construction industry, rural housing and biomedical applications.

## ACKNOWLEDGEMENT

The author, Mohammad Jawaid would like to thank Universiti Teknologi Malaysia for the Visiting Lecturer Position that has made this research work possible.

## REFERENCE

1. Khalil. H. P. S. Abdul, A. H. Bhat, M. Jawaid, P. Amouzgar, R. Ridzuan, M. R. Said, 2010aAgro-Wastes: Mechanical and physical properties of resin impregnated oil palm trunk core lumber. Polymer Composites 314638644
2. Khalil. H. P. S. Abdul, M. Y. Nur, M. Firdaus, M. Jawaid, R. Anis, Ridzuan, A. R. Mohamed, 2010bDevelopment and material properties of new hybrid medium density fibreboard from empty fruit bunch and rubberwood. Materials & Design 31942294236

3. Khalil. H. P. S. Abdul, S. Hanida, C. W. Kang, N. A. Nik, Fuaad, 2007Agro-hybrid composite: The effects on mechanical and physical properties of oil palm fiber (EFB)/glass hybrid reinforced polyester composites. Journal of Reinforced Plastics and Composites 262203218
4. Khalil. H. P. S. Abdul, M. Jawaid, A. Abu, Bakar, 2011Woven hybrid composites: Water absorption and thickness swelling behaviours. BioResources 6210431052
5. Khalil. H. P. S. Abdul, C. W. Kang, A. Khairul, R. Ridzuan, T. O. Adawi, 2009The effect of different laminations on mechanical and physical properties of hybrid composites. Journal of Reinforced Plastics and Composites 28911231137
6. Khalil. H. P. S. Abdul, M. Y. Nur, M. Firdaus, Anis, R. Ridzuan, 2008The Effect of Storage Time and Humidity on Mechanical and Physical Properties of Medium Density Fiberboard (MDF) from Oil Palm Empty Fruit Bunch and Rubberwood. Polymer-Plastics Technology and Engineering 471010461053
7. Khalil. H. P. S. Abdul, M. R. Nurul, A. H. Fazita, M. Bhat, Jawaid, N. A. Nik, Fuad, 2010cDevelopment and material properties of new hybrid plywood from oil palm biomass. Materials & Design 311417424
8. Khalil. H. P. S. Abdul, B. T. Poh, A. M. Issam, M. Jawaid, R. Ridzuan, 2010dRecycled polypropylene-oil palm biomass: The effect on mechanical and physical properties. Journal of Reinforced Plastics and Composites 29811171130
9. Khalil. H. P. S. Abdul, H. D. Rozman, 2004Gentian dan lignoselulosik: Universiti Sains Malaysia: Pulau Pinang.
10. Khalil. H. P. S. Abdul, Alwani. M. Siti, Omar. A. Mohd, K., 2006Chemical composition, anatomy, lignin distribution, and cell wall structure of Malaysian plant fibers. BioResources 12220232
11. Bakar. A. Abu, Hariharan, H. P. S. Abdul, Khalil, 2005Lignocellulose-based hybrid bilayer laminate composite: Part I- Studies on tensile and impact behavior of oil palm fiber-glass fiber-reinforced epoxy resin. Journal of Composite Materials 398663684
12. Hassan. O. Abu, M. Ishida, Shukri. I. Mohd, Tajuddin. Ahmad, Z., 1994Oil-Palm Fronds as a Roughage Feed Source for Ruminants in Malaysia. Malaysia Agriculture Research and Development Institute (MARDI), Kuala Lumpur, Malaysia.
13. R. Agarwal, N. S. Saxena, K. B. Sharma, S. Thomas, L. A. Pothan, 2003Thermal conduction and diffusion through glass-banana fiber polyester composites. Indian Journal of Pure and Applied Physics 41 (6 SPEC.):448452

14. R. Agrawal, N. S. Saxena, M. S. Sreekala, S. Thomas, 2000Non-isothermal crystallization kinetics of glass oil palm fibre reinforced phenolformaldehyde composites. Journal of Scientific and Industrial Research 592136139

15. Richa. N. S. Agrawal, M. S. Saxena, Sreekala, S. Thomas, 2000Effect of treatment on the thermal conductivity and thermal diffusivity of oil-palm-fiber-reinforced phenolformaldehyde composites. Journal of Polymer Science Part B: Polymer Physics 387916921

16. Z. Ahmad, H. M. Saman, P. M. Tahir, 2010Oil palm trunk fibre as a bio-waste resource for concrete reinforcement. International Journal of Mechanical and Materials Engineering 52199207

17. I. S. Aji, S. M. Sapuan, E. S. Zainudin, K. Abdan, 2009Kenaf fibres as reinforcement for polymeric composites: A review. International Journal of Mechanical and Materials Engineering 43239248

18. H. M. Akil, L. W. Cheng, Z. A. Mohd, A. Ishak, Bakar. Abu, M. A. Abd, Rahman, 2009Water absorption study on pultruded jute fibre reinforced unsaturated polyester composites. Composites Science and Technology 69 (11-12):1942-1948.

19. H. A. Al-Qureshi, 2001The application of jute fibre reinforced composites for the development of a car body. Paper read at UMIST Conference, at UK.

20. Z. Alam, A. A. Mamun, I. Y. Qudsieh, S. A. Muyibi, H. M. Salleh, N. M. Omar, 2009Solid state bioconversion of oil palm empty fruit bunches for cellulase enzyme production using a rotary drum bioreactor. Biochemical Engineering Journal 4616164

21. K. M. Amar, M. Manjusri, T. D. Lawrence, ed, 2005Natural Fibers, Biopolymers, and Biocomposites. USA: CRC Press, Tayler and Francis Group.

22. H. Anuar, S. H. Ahmad, R. Rasid, N. S. Nik, Daud, 2006Tensile and impact properties of thermoplastic natural rubber reinforced short glass fiber and empty fruit bunch hybrid composites. Polymer-Plastics Technology and Engineering 45910591063

23. M. F. Ashby, Y. J. M. Brechet, 2003Designing hybrid materials. Acta Materialia 511958015821

24. A. A. Bakar, A. Hassan, A. F. Mohd, Yusof, 2006The effect of oil extraction of the oil palm empty fruit bunch on the processability, impact, and flexural properties of PVC-U composites. International Journal of Polymeric Materials 559627641

25. Y. Basiron, 2007Palm oil production through sustainable plantations. European Journal of Lipid Science and Technology 1094289295

26. Y. Basiron, M. A. Simeh, 2005Vision 2020- The oil palm phenonmenon. Palm Industry Economic Journal. 52110

27. G. Basu, A. N. Roy, 2007Blending of jute with different natural fibres. Journal of Natural Fibers 441329

28. Amor. I. Ben, H. Rekik, H. Kaddami, M. Raihane, M. Arous, A. Kallel, 2010Effect of palm tree fiber orientation on electrical properties of palm tree fiber-reinforced polyester composites. Journal of Composite Materials 441315531568

29. A. Bismarck, S. Mishra, T.. Lampke, 2005Plant fibres as reinforcement for green composites. In Natural fibres biopolymers and biocomposites, edited by A. K. Mohanthy, Misra, M., Drzal, L.T. USA: CRC Press, Taylor and Francis Group.

30. A. K. Bledzki, O. Faruk, V. E. Sperber, 2006Cars from bio fibres. Macromolecular Materials and Engineering 291449457

31. A. K. Bledzki, J. Gassan, 1999Composites reinforced with cellulose based fibres. Progress in Polymer Science (Oxford) 242221274

32. L. L. Chai, S. Zakaria, C. H. Chia, S. Nabihah, R. Rasid, 2009Physico-mechanical properties of PF composite board from EFB fibres using liquefaction technique. Iranian Polymer Journal (English Edition) 1811917923

33. N. Chand, 1992Electrical characteristics of sunhemp fibre. Journal of Materials Science Letters 113138139

34. N. Chand, D. Jain, 2005Effect of sisal fibre orientation on electrical properties of sisal fibre reinforced epoxy composites. Composites Part A: Applied Science and Manufacturing 365594602

35. L. T. Chew, C. L. Ong, 1985Particleboard from oil palm trunk. Paper read at Proceedings of the National Symposium on Oil Palm By-products for Agro-based Industries, at Kaula Lumpur.

36. T. L. Chew, S. Bhatia, 2008Catalytic processes towards the production of biofuels in a palm oil and oil palm biomass-based biorefinery. Bioresource Technology 991779117922

37. I. Dahlan, 2000Oil palm frond, a feed for herbivores. Asian-Australasian Journal of Animal Sciences 13 (SUPPL. C):300303

38. A. K. Datta, B. K. Samantaray, S. Bhattacherjee, 1984Mechanical and dielectric properties of pineapple fibres. Journal of Materials Science Letters 38667670

39. H. Demir, U. Atikler, D. Balköse, F. Tihm, Ioglu. Inl, 2006The effect of fiber surface treatments on the tensile and water sorption properties of polypropylene-luffa fiber composites. Composites Part A: Applied Science and Manufacturing 373447456

40. A. Demirbaş, 2000Mechanisms of liquefaction and pyrolysis reactions of biomass. Energy Conversion and Management 416633646
41. A. Dufresne, 2008Cellulose-based composites and nanocomposites. In Monomers, Polymers and Composites from Renewable Resources, edited by A. Gandini, Belgacem, M.N. Oxford, UK: Elsevier.
42. John. Duncan, 2008Principle and applications of mechanical thermal analysis. In Principles and Applications of Thermal Analysis edited by P. Gabbott. Oxford: Blackwell Publishing Ltd, UK.
43. A. N. Fraga, E. Frullloni, O. De La Osa, J. M. Kenny, A. VaÌ□zquez, 2006Relationship between water absorption and dielectric behaviour of natural fibre composite materials. Polymer Testing 252181187
44. S. Y. Fu, G. Xu, Y. W. Mai, 2002On the elastic modulus of hybrid particle/short-fiber/polymer composites. Composites Part B:Engineering 334291299
45. A. K. Gupta, 2009Natural plant fibre based green composite for automobile application.
46. A. Hariharan, Bakar. Abu, Khalil. H. P. S. Abdul, 2004Influence of Oil Palm fibre loading on the mechanical and physical properties of glass fibre reinforced epoxy bi-layer hybrid laminated composite. Paper read at Proceeding of 3rd USM-JIRCAS Joint International Symposium, 911March, 2004, at Penang, Malaysia.
47. A. P. Haseena, G. Unnikrishnan, G. Kalaprasad, 2007Dielectric properties of short sisal/coir hybrid fibre reinforced natural rubber composites. Composite Interfaces 14 (7-9):763-786.
48. Hassan, Azman, Arshad Adam Salema, Farid Nasir Ani, and Aznizam Abu Bakar.2010A review on oil palm empty fruit bunch fiber-reinforced polymer composite materials. Polymer Composites 311220792101
49. H. A. Hassim, M. Lourenço, G. Goel, B. Vlaeminck, Y. M. Goh, V. Fievez, 2010Effect of different inclusion levels of oil palm fronds on in vitro rumen fermentation pattern, fatty acid metabolism and apparent biohydrogenation of linoleic and linolenic acid. Animal Feed Science and Technology 162 (3-4):155-158.
50. C. A. S. Hill, H. P. S. Abdul, Khalil, 2000Effect of fiber treatments on mechanical properties of coir or oil palm fiber reinforced polyester composites. Journal of Applied Polymer Science 78916851697
51. M. Idicula, S. K. Malhotra, K. Joseph, S. Thomas, 2005aDynamic mechanical analysis of randomly oriented intimately mixed short banana/sisal hybrid fibre reinforced polyester composites. Composites Science and Technology 65 (7-8):1077-1087.

52. M. Idicula, S. K. Malhotra, K. Joseph, S. Thomas, 2005bEffect of layering pattern on dynamic mechanical properties of randomly oriented short banana/sisal hybrid fiber-reinforced polyester composites. Journal of Applied Polymer Science 97521682174

53. M. Jacob, B. Francis, S. Thomas, K. T. Varughese, 2006cDynamical mechanical analysis of sisal/oil palm hybrid fiber-reinforced natural rubber composites. Polymer Composites 276671680

54. M. Jacob, B. Francis, K. T. Varughese, S. Thomas, 2006dThe effect of silane coupling agents on the viscoelastic properties of rubber biocomposites. Macromolecular Materials and Engineering 291911191126

55. M. Jacob, S. Jose, S. Thomas, K. T. Varughese, 2006bStress relaxation and thermal analysis of hybrid biofiber reinforced rubber biocomposites. Journal of Reinforced Plastics and Composites 251819031917

56. M. Jacob, S. Thomas, K. T. Varughese, 2004aNatural rubber composites reinforced with sisal/oil palm hybrid fibers: Tensile and cure characteristics. Journal of Applied Polymer Science 93523052312

57. M. Jacob, S. Thomas, K. T. Varughese, 2004bMechanical properties of sisal/oil palm hybrid fiber reinforced natural rubber composites. Composites Science and Technology 64 (7-8):955-965.

58. M. Jacob, K. T. Varughese, S. Thomas, 2005Water sorption studies of hybrid biofiber-reinforced natural rubber biocomposites. Biomacromolecules 6629692979

59. M. Jacob, K. T. Varughese, S. Thomas, 2006aDielectric characteristics of sisal-oil palm hybrid biofibre reinforced natural rubber biocomposites. Journal of Materials Science 411755385547

60. Maya. Jacob, Sabu, Thomas, K. T. Varughese, 2007Biodegradability and Aging Studies of Hybrid Biofiber Reinforced Natural Rubber Biocomposites. Journal of Biobased Materials and Bioenergy 1118126

61. K. Jarukumjorn, N. Suppakarn, 2009Effect of glass fiber hybridization on properties of sisal fiber-polypropylene composites. Composites Part B: Engineering 407623627

62. M. Jawaid, H. P. S. Abdul, Khalil, 2011aCellulosic/synthetic fibre reinforced polymer hybrid composites: A review. Carbohydrate Polymers 861118

63. M. Jawaid, H. P. S. Abdul, Khalil, A. Abu, Bakar, 2010Mechanical performance of oil palm empty fruit bunches/jute fibres reinforced epoxy hybrid composites. Materials Science and Engineering A 527 (29-30):7944-7949.

64. M. Jawaid, H. P. S. Abdul, Khalil, A. Abu, Bakar, 2011bWoven hybrid

composites: Tensile and flexural properties of oil palm-woven jute fibres based epoxy composites. Materials Science and Engineering A 5281551905195

65. M. Jawaid, H. P. S. Abdul, Khalil, O. S. Alattas, 2012Woven hybrid biocomposites: Dynamic mechanical and thermal properties. Composites Part A: Applied Science and Manufacturing 432288293

66. M. Jawaid, H. P. S. Abdul, A. H. Khalil, Bhat, A. Abu, Baker, 2011cImpact properties of natural fiber hybrid reinforced epoxy composites. Advanced Materials Research 264-265:688-693.

67. M. Jawaid, H. P. S. Abdul, P. Khalil, Khanam. Noorunnisa, A. Abu, Bakar, 2011dHybrid Composites Made from Oil Palm Empty Fruit Bunches/Jute Fibres: Water Absorption, Thickness Swelling and Density Behaviours. Journal of Polymers and the Environment 191106109

68. M. Jawaid, Khalil. H. P. S. Abdul, 2011eEffect of layering pattern on the dynamic mechanical properties and thermal degradation of oil palm-jute fibers reinforced epoxy hybrid composite. BioResources 6323092322

69. M. Jawaid, H. P. S. A. Khalil, A. A. Bakar, 2011fHybrid composites of oil palm empty fruit bunches/woven jute fiber: Chemical resistance, physical, and impact properties. Journal of Composite Materials 452425152522

70. M. Jawaid, H. P. S. A. Khalil, A. A. Bakar, P. N. Khanam, 2011gChemical resistance, void content and tensile properties of oil palm/jute fibre reinforced polymer hybrid composites. Materials and Design 32210141019

71. G. Jayamol, S. S. Bhagawan, S. Thomas, 1998Electrical properties of pineapple fibre reinforced polyethylene composites. Journal of Polymer Engineering 175383404

72. M. J. John, B. Francis, K. T. Varughese, S. Thomas, 2008Effect of chemical modification on properties of hybrid fiber biocomposites. Composites Part A: Applied Science and Manufacturing 392352363

73. M. J. John, S. Thomas, 2008Biofibres and biocomposites. Carbohydrate Polymers 713343364

74. Jon Fox-Rubin, David Cramer.2010Thermoplastic Composites. Flex Technologies 2010 [cited 29Th October 2010]. Available from http://www.fiberforge.com/thermoplastic-composites/thermoplastic-composites.php.

75. S. Joseph, S. Thomas, 2008Electrical properties of banana fiber-reinforced phenol formaldehyde composites. Journal of Applied

Polymer Science 1091256263

76. S. V. Joshi, L. T. Drzal, A. K. Mohanty, S. Arora, 2004Are natural fiber composites environmentally superior to glass fiber reinforced composites? Composites Part A: Applied Science and Manufacturing 353371376
77. A. Kalam, B. B. Sahari, Y. A. Khalid, S. V. Wong, 2005Fatigue behaviour of oil palm fruit bunch fibre/epoxy and carbon fibre/ epoxy composites. Composite Structures 7113444
78. J. Kaliwon, Ahmad. S. Sh, Aziz. A. Abdul, 2010International Conference on Science and Social Research (CSSR). 57Dec. 2010. 528-531. Kuala Lumpur, Malaysia
79. J. Karger-Kocsis, 2000Reinforced polymer blends. In Polymer Blends, edited by D. R. Paul, Bucknall, C.B.,. New York: John Wiley & Sons.
80. M. Karina, H. Onggo, A. H. Dawam, Abdullah, A. Syampurwadi, 2008Effect of oil palm empty fruit bunch fiber on the physical and mechanical properties of fiber glass reinforced polyester resin. Journal of Biological Sciences 81101106
81. H. P. S. A. Khalil, H. D. Rozman, M. N. Ahmad, H. Ismail, 2000Acetylated plant-fiber-reinforced polyester composites: A study of mechanical, hygrothermal, and aging characteristics. Polymer-Plastics Technology and Engineering 394757781
82. H. P. S. Khalil, M. Abdul, Alwani. R. Siti, H. Ridzuan, Kamarudin, A. Khairul, 2008Chemical Composition, Morphological Characteristics, and Cell Wall Structure of Malaysian Oil Palm Fibers. Polymer-Plastics Technology and Engineering 473273280
83. K. C. Khoo, T. W. Lee, 1985Sulphate pulping of the oil palm trunk. Paper read at National Symposium on Oil Palm By-Products for Agro-Based Industries, at Kaulalumpur, Malaysia.
84. G. Kickelbick, ed, 2007Introduction to Hybrid Materials. Edited by G. Kickelbick, Hybrid Materials:Synthesis, Characterization, and Applications Weinheim: WILEY-VCH Verlag GmbH & Co. KGaA.
85. W. Killman, L. T. Hong, 1989Anatomy and properties of oil palm stem.Proceeding of the national symposium of oil palm by-products for agro-based industries, Kaula Lumpur, 1842
86. A. G. Kulkarni, K. G. Satyanarayana, P. K. Rohatgi, 1981Electrical resistivity of some plant fibers. J. Mater. Sci. 161719
87. R. N. Kumar, L. M. Wei, H. D. Rozman, A. Abusamah, 1997Fire resistant sheet moulding composites from hybrid reinforcements of oil palm-fibres and glass fibre. International Journal of Polymeric Materials 37 (1-2):43-52.

88. K. N. Law, W. R. W. Daud, A. Ghazali, 2007Morphological and chemical nature of fiber strands of oil palm empty-fruit-bunch (OPEFB). BioResources 23351362
89. Law, Kwei-Nam, and Xingfen Jiang.2001Comparative papermaking properties of oil-palm empty fruit bunch. TAPPI Journal 84 (1):95.
90. S. H. Lee, S. Wang, 2006Biodegradable polymers/bamboo fiber biocomposite with bio-based coupling agent. Composites Part A: Applied Science and Manufacturing 3718091
91. K. O. Lim, 1998Oil palm plantations- A plausible renewable source of energy. International Energy Journal 202107116
92. S. C. Lim, K. Khoo, C.198, K.C.1986Characterization of oil palm trunk and its potential utilization.Malaysian Forester 49322
93. M. Marzinotto, C. Santulli, C. Mazzetti, 2007Dielectric properties of oil palm-natural rubber biocomposites.Conference on Electrical insulation and Dielectric Phenomena, CEIDP 2007.
94. Mat Amin, Khairul Anuar, and Khairiah Haji Badri.2007Palm-based bio-composites hybridized with kaolinite. Journal of Applied Polymer Science 105524882496
95. N. M. Mehta, P. H. Parsania, 2006Fabrication and evaluation of some mechanical and electrical properties of jute-biomass based hybrid composites. Journal of Applied Polymer Science 100317541758
96. D. Meier, O. Faix, 1999State of the art of applied fast pyrolysis of lignocellulosic materials- A review. Bioresource Technology 6817177
97. Tomoaki. Minowa, Kondo. Teruo, T. Soetrisno, Sudirjo, 1998Thermochemical liquefaction of indonesian biomass residues. Biomass and Bioenergy 14 (5-6):517-524.
98. Supriya. Mishra, K. Amar, Lawrence. T. Mohanty, Manjusri. Drzal, Misra, Hinrichsen. Georg, 2004A Review on Pineapple Leaf Fibers, Sisal Fibers and Their Biocomposites. Macromolecular Materials and Engineering 28911955974
99. H. Mohamad, Zawawi. Z. Zin, H. Abdul, Halim, 1985Potentials of oil palm by-products as raw materials for agro-based industries. Paper read at National Symposium on Oil Palm By-Products for Agro-Based Industries., at Kaulalumpur, Malaysia.
100. MPOB.Overview of the Malaysian Oil Palm Industry 2009MPOB 2009.
101. S. Mridha, S. B. Keng, Z. Ahmad, 2007The effect of OPWF filler on impact strength of glass-fiber reinforced epoxy composite. Journal of Mechanical Science and Technology 211016631670
102. Gowda. T. Munikenche, A. C. B. Naidu, R. Chhaya, 1999Some mechanical properties of untreated jute fabric-reinforced polyester

composites. Composites Part A: Applied Science and Manufacturing 303277284

103. A. B. Nasrin, A. N. Y. M. Choo, S. Mohamad, M. H. Rohaya, A. Azali, Z. Zainal, 2008Oil palm biomass as potential substitution raw materials for commercial biomass Briquettes production. American Journal of Applied Sciences 53179183

104. S. K. Nayak, S. Mohanty, S. K. Samal, 2010aInfluence of interfacial adhesion on the structural and mechanical behavior of PP-banana/ glass hybrid composites. Polymer Composites 31712471257

105. S. K. Nayak, S. Mohanty, 2010bSisal glass fiber reinforced PP hybrid composites: Effect of MAPP on the dynamic mechanical and thermal properties. Journal of Reinforced Plastics and Composites 291015511568

106. S. K. Nayak, S. Mohanty, S. K. Samal, 2009Influence of short bamboo/ glass fiber on the thermal, dynamic mechanical and rheological properties of polypropylene hybrid composites. Materials Science and Engineering A 523 (1-2):32-38.

107. S. K. Nayak, S. Mohanty, S. K. Samal, 2010cHybridization effect of glass fibre on mechanical, morphological and thermal properties of polypropylene-bamboo/glass fibre hybrid composites. Polymers and Polymer Composites 184205218

108. Haliza. A. H. Noor, A. Fazilah, M. Azemi, N., 2006Development of hemicelluloses biodegradable films from oil palm from (Elais Guinnesis). In International Conference on Green and Sustainable Innovation. Chang Mai, Thailand: Energy Management and Conservation Centre, Chang Mai university.

109. Khanam. P. Noorunnisa, H. P. S. Abdul, M. Khalil, G. Jawaid, Reddy. C. Ramachandra, Narayana. Surya, S. Venkata, Naidu, 2010Sisal/ Carbon Fibre Reinforced Hybrid Composites: Tensile, Flexural and Chemical Resistance Properties. Journal of Polymers and the Environment:17

110. K. Oksman, 2000Mechanical properties of natural fibre mat reinforced thermoplastic. Applied Composite Materials 7 (5-6):403-414.

111. H. L. Ornaghi Jr, A. S. Bolner, R. Fiorio, A. J. Zattera, S. C. Amico, 2010Mechanical and dynamic mechanical analysis of hybrid composites molded by resin transfer molding. Journal of Applied Polymer Science 1182887896

112. S. Panigrahi, R. L. Kushwaha, Sujata, Panigrahi.2010Characterization of Palm Fiber for Development of Biocomposites Material for Automotive Industries.SAE International Journal of Commercial Vehicles 31304312

113. S. Panthapulakkal, M. Sain, 2007Injection-molded short hemp fiber/ glass fiber-reinforced polypropylene hybrid composites-mechanical, water absorption and thermal properties. Journal of Applied Polymer Science 103424322441

114. A. Paul, S. Thomas, 1997Electrical properties of natural-fiber-reinforced low density polyethylene composites: A comparison with carbon black and glass-fiber-filled low density polyethylene composites. Journal of Applied Polymer Science 632247266

115. L. A. Pothan, C. N. George, M. J. John, S. Thomas, 2010Dynamic mechanical and dielectric behavior of banana-glass hybrid fiber reinforced polyester composites. Journal of Reinforced Plastics and Composites 29811311145

116. P. Punsuvon, W. Anpanurak, P. Vaithanomsat, N.. Tungkananuruk, 2005Fractionation of chemical components of oil palm trunk by steam explosion. Paper read at 31st Congress on Science and Technology of Thailand, at Suranaree University of Technology, Thailand.

117. A. K. Rana, A. Mandal, S. Bandyopadhyay, 2003Short jute fiber reinforced polypropylene composites: effect of compatibiliser, impact modifier and fiber loading. Composites Science and Technology 636801806

118. Jegatheswaran. Ratnasingam, Oil Palm Biomass Utilization- Counting the Successes in Malaysia. 2011Available from http://www.woodmagmagazine.com/node/417.

119. K. Raveendran, A. Ganesh, K. C. Khilar, 1995Influence of mineral matter on biomass pyrolysis characteristics. Fuel 741218121822

120. Reddy, Narendra, and Yiqi Yang.2005Biofibers from agricultural byproducts for industrial applications. Trends in Biotechnology 2312227

121. Jonathan. D. Reid, H. William, Lawrence, P. Richard, Buck, 1986Dielectric Properties Of An Epoxy Resin And Its Composite. I. Moisture Effects On Dipole Relaxation. Journal of Applied Polymer Science 31617711784

122. J. Rout, M. Misra, S. S. Tripathy, S. K. Nayak, A. K. Mohanty, 2001The influence of fibre treatment of the performance of coir-polyester composites. Composites Science and Technology 61913031310

123. H. D. Rozman, L. Musa, A. Abubakar, 2005Rice husk-polyester composites: The effect of chemical modification of rice husk on the mechanical and dimensional stability properties. Journal of Applied Polymer Science 97312371247

124. H. D. Rozman, G. S. Tay, R. N. Kumar, A. Abusamah, H. Ismail, Z.

A. Mohd, 2001bThe effect of oil extraction of the oil palm empty fruit bunch on the mechanical properties of polypropylene-oil palm empty fruit bunch-glass fibre hybrid composites. Polymer-Plastics Technology and Engineering 402103115

125. H. D. Rozman, G. S. Tay, R. N. Kumar, A. Abusamah, H. Ismail, Z. A. Mohd, Ishak, 2001aPolypropylene-oil palm empty fruit bunch-glass fibre hybrid composites: A preliminary study on the flexural and tensile properties. European Polymer Journal 37612831291
126. S.Yacob.2007Progress and challenges in utilization of oil palm biomass. In Asian Science and Technology Seminar. Jakarta, Indonesia.
127. S. K. Samal, S. Mohanty, S. K. Nayak, 2009aBanana/glass fiber-reinforced polypropylene hybrid composites: Fabrication and performance evaluation. Polymer- Plastics Technology and Engineering 484397414
128. S. K. Samal, S. Mohanty, S. K. Nayak, 2009bPolypropylene-bamboo/glass fiber hybrid composites: Fabrication and analysis of mechanical, morphological, thermal, and dynamic mechanical behavior. Journal of Reinforced Plastics and Composites 282227292747
129. K. G. Satyanarayana, C. K. S. Pillai, K. Sukumaran, S. G. K. Pillai, P. K. Rohatgi, K. Vijayan, 1982Structure property studies of fibres from various parts of the coconut tree. Journal of Materials Science 17824532462
130. T. G. Schuh, 201, Renewable materials for automotive applications 2004cited 29.09.2010]. Available from www.ienica.net/fibresseminar/schuh.pdf.
131. Sheffield University.2011Biodegradable cars in the future? http://www.cars-tips.com/best-car-tips/biodegradable-cars-in-the-future/
132. S. Shinoj, R. Visvanathan, S. Panigrahi, 2010Towards industrial utilization of oil palm fibre: Physical and dielectric characterization of linear low density polyethylene composites and comparison with other fibre sources. Biosystems Engineering 1064378388
133. S. Shinoj, R. Visvanathan, S. Panigrahi, M. Kochubabu, 2011Oil palm fiber (OPF) and its composites: A review. Industrial Crops and Products 331722
134. S. H. Shuit, K. T. Tan, K. T. Lee, A. H. Kamaruddin, 2009Oil palm biomass as a sustainable energy source: A Malaysian case study. Energy 34912251235
135. Son, Jungil, Hyun-Joong Kim, and Phil-Woo Lee.2001Role of paper sludge particle size and extrusion temperature on performance of paper sludge-thermoplastic polymer composites. Journal of Applied

Polymer Science 821127092718

136. M. S. Sreekala, J. George, M. G. Kumaran, S. Thomas, 2001Water-sorption kinetics in oil palm fibers. Journal of Polymer Science, Part B: Polymer Physics 391112151223

137. M. S. Sreekala, J. George, M. G. Kumaran, S. Thomas, 2002aThe mechanical performance of hybrid phenol-formaldehyde-based composites reinforced with glass and oil palm fibres. Composites Science and Technology 623339353

138. M. S. Sreekala, M. G. Kumaran, M. L. Geethakumariamma, S. Thomas, 2004Environmental effects in oil palm fiber reinforced phenol formaldehyde composites: Studies on thermal, biological, moisture and high energy radiation effects. Advanced Composite Materials: The Official Journal of the Japan Society of Composite Materials 13 (3-4):171-197.

139. M. S. Sreekala, M. G. Kumaran, S. Thomas, 1997Oil palm fibers: Morphology, chemical composition, surface modification, and mechanical properties. Journal of Applied Polymer Science 665821835

140. M. S. Sreekala, M. G. Kumaran, S. Thomas, 2002bWater sorption in oil palm fiber reinforced phenol formaldehyde composites. Composites-Part A: Applied Science and Manufacturing 336763777

141. M. S. Sreekala, S. Thomas, G. Groeninckx, 2005Dynamic mechanical properties of oil palm fiber/phenol formaldehyde and oil palm fiber/glass hybrid phenol formaldehyde composites. Polymer Composites 263388400

142. F. Sulaiman, N. Abdullah, H. Gerhauser, A. Shariff, 2011An outlook of Malaysian energy, oil palm industry and its utilization of wastes as useful resources. Biomass and Bioenergy 35937753786

143. S. A. Sulaiman, M. R. T. Ahmad, S. Atnaw, M., 2011Prediction of Biomass Conversion Process for Oil Palm Fronds in a Downdraft Gasifier. Paper read at The 4th International Meeting of Advances in Thermofluids, 34th October, 2011, at Melaka, Malaysia.

144. S. Sumathi, S. P. Chai, A. R. Mohamed, 2008Utilization of oil palm as a source of renewable energy in Malaysia. Renewable and Sustainable Energy Reviews 12924042421

145. G. Swaminathan, K. Shivakumar, 2009A Re-examination of DMA testing of polymer matrix composites. Journal of Reinforced Plastics and Composites 288979994

146. M. M. Thwe, K. Liao, 2003Durability of bamboo-glass fiber reinforced polymer matrix hybrid composites. Composites Science and Technology 63 (3-4):375-387.

147. Devi. L. Uma, S. S. Bhagawan, S. Thomas, 2010Dynamic mechanical analysis of pineapple leaf/glass hybrid fiber reinforced polyester composites. Polymer Composites 316956965

148. K. Van Rijswijk, W. D. Brouwer, A. Beukers, 2001Application of Natural Fibre Composites in the Development of Rural Societies. edited by A. Engineering: FAO.

149. Edmund. Viva, 2011Palm oil waste to be processed into composite board. http://www.varieart.com/article137 -Palm-Oil-Waste-to-be-Processed-into-Composite-BoardAccessed on 13th April 2012]

150. Busu. W. N. Wan, H. Anuar, S. H. Ahmad, R. Rasid, N. A. Jamal, 2010The mechanical and physical properties of thermoplastic natural rubber hybrid composites reinforced with Hibiscus cannabinus, L and short glass fiber. Polymer- Plastics Technology and Engineering 491313151322

151. Nadirah. W. O. Wan, M. Jawaid, A. A. Al, H. P. S. Masri, Khalil. S. S. Abdul, Suhaily, A. R. Mohamed, 2011Cell Wall Morphology, Chemical and Thermal Analysis of Cultivated Pineapple Leaf Fibres for Industrial Applications. Journal of Polymers and the Environment:18

152. Rosli. W. D. Wan, K. N. Law, Z. Zainuddin, R. Asro, 2004Effect of pulping variables on the characteristics of oil-palm frond-fiber. Bioresource Technology 933233240

153. K. J. Wong, U. Nirmal, B. K. Lim, 2010Impact behavior of short and continuous fiber-reinforced polyester composites. Journal of Reinforced Plastics and Composites 292334633474

154. B. Xiao, X. F. Sun, R. Sun, 2001Chemical modification of lignins with succinic anhydride in aqueous systems. Polymer Degradation and Stability 712223231

155. J. M. Yatim, Khalid. N. H. Abd, Reza, 2011Seminar Embracing Green Technology In Construction- Way Forward., 26th April, Grand Margherita Hotel, Kuching Sarawak, Malaysia

156. A. Zaidon, A. M. Norhairul, M. Y. Nizam, Nor. F. Mohd, M. T. Abood, M. Y. Paridah, Yuziah. Nor, H. Jalaluddin, 2007Properties of particleboard made from pretreated particles of rubberwood, EFB and rubberwood-EFB blend. Journal of Applied Sciences 7811451151

157. R. Zulkifli, M. J. M. Nor, A. R. Ismail, M. Z. Nuawi, S. Abdullah, M. F. M. Tahir, M. N. A. Rahman, 2009Comparison of acoustic properties between coir fibre and oil palm fibre. European Journal of Scientific Research 331144152

# Chapter 10

# CARBON FIBER SENSOR: THEORY AND APPLICATION

Alexander Horoschenkoff[1] and Christian Christner[2]

[1] Munich University of Applied Sciences, Germany

[2] Universität der Bundeswehr München, Germany

## INTRODUCTION

The piezoresistive[1] - carbon fiber sensor (CFS) consists of a single carbon fiber roving with electrical connected endings embedded in a sensor carrier (patch) for electrical insulation. Depending on the requirements of the application different patch types (e.g. glass fiber reinforced plastic (GFRP), polyester film, neat epoxy resin) are used. In terms of the mechanical properties GFRP is a particularly suitable patch material. CFSs with a GFRP patch exhibits an improved linearity of the signal due to the supportive effect of the glass fibers to the carbon sensor fiber especially in the case of compression loading. In [6] the ex-PAN fiber T300B was identified as one suitable carbon fiber for CFSs because of the excellent linear piezoresistive behavior, the high specific resistivity and a high breaking elongation of the fiber. Figure 1 shows a CFS with a single layer GFRP patch (UD prepreg EG/913).

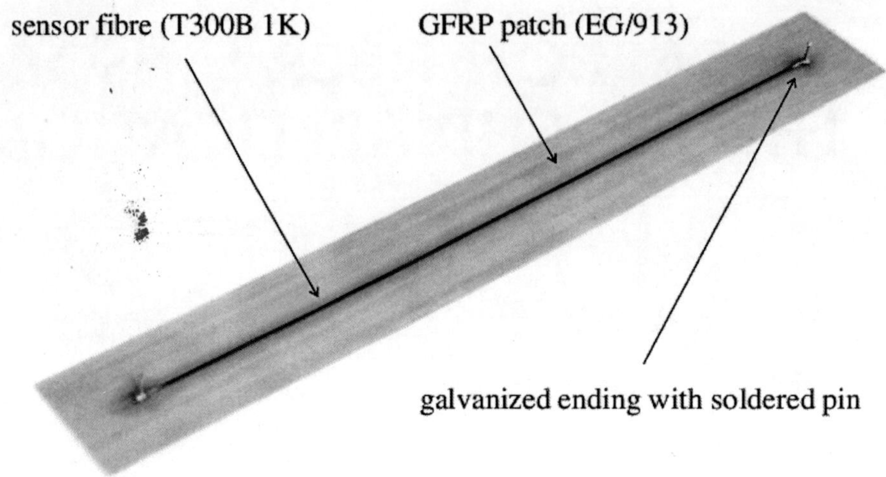


**Figure 1.** Carbon fiber sensor with an ex-Pan sensor fiber and a GFRP patch.

The manufacturing process of a CFS includes three basic steps: Pre-curing of the carbon fiber, preparation of the electrical connection and embedding of the sensor fiber into the sensor carrier.

The pre-curing process is used to stabilize the carbon fiber roving and to align the filaments of the roving. For this purpose the twisted carbon fiber roving is impregnated by a resin with low viscosity and cured by using a special tooling. Good results for the impregnation of the carbon fiber roving T300B 1K were obtained by using the epoxy resin EP301 S (HBM) and a twist of 20 turns per meter. Spring elements provided a constant tension force along the roving during the curing process at 180°C for 1.5 hours.

For preparation of electrical connections a galvanic process is applied based on a nickel electrolyte. In order to attain a homogeneous nickel coating of the filaments the resin must be removed at the fiber endings[2] - . An applied current of 40 mA for 30 seconds leads to an excellent nickel coating. Depending on the application of the CFS (surface application or structural integration) the ends of the sensor fiber can be provided with soldered pins.

The embedding process of the sensor fiber depends on the used patch type and patch material. Figure 2 shows a micro section of a carbon fiber sensor in longitudinal and transverse direction.

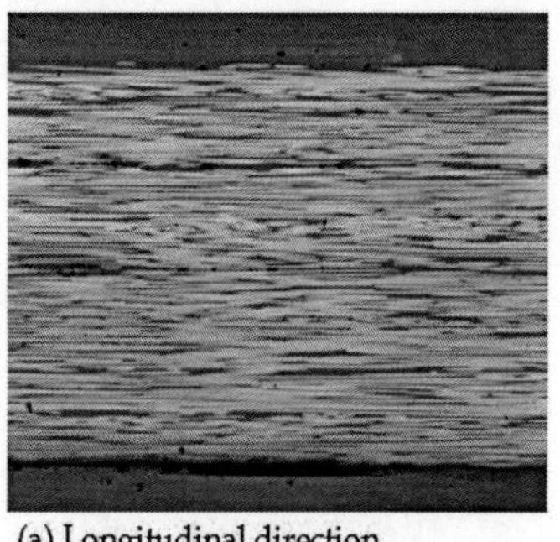

(a) Longitudinal direction (b) Transverse direction

**Figure 2.** Microsection of a carbon fiber sensor in longitudinal and transverse direction Sensorfiber: 1K roving of the carbon fiber T300B Patch: single layer UD prepreg EG/913

## ELECTROMECHANICAL PROPERTIES OF CARBON FIBERS

Referring to Chung [2] and Dresselhaus [3] crystalline carbon fibers have the same crystal structure as graphite. The layered and planar structure of crystalline carbon fibers is shown in Figure 3. In each layer the $sp^2$ hybridized carbon atoms are arranged in a hexagonal lattice. Within a layer (x-y plane) the carbon atoms are bonded by three covalent bonds (overlapping $sp^2$ orbitals), and a metallic bonding is provided by the delocalization of the $p_z$ orbitals. This delocalization of the fourth valence electron leads to a good electrical conductivity of the carbon fiber. The individual layers are held together by weak van der Waals forces (z-direction). These different types of bonds within and between the layers result in the anisotropy of the mechanical, thermal and electrical material properties of carbon fibers.

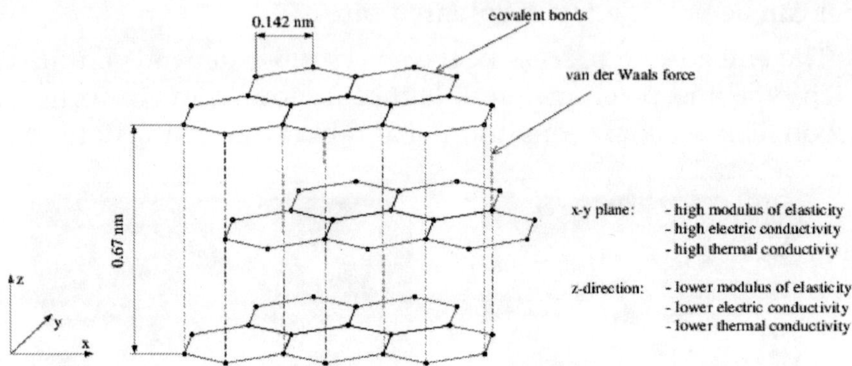


**Figure 3.** Graphite structure of crystalline carbon fibers.

## Specific Resistivity

The degree of crystallinity and the micro structure of carbon fibers are controlled by the carbonization process and determine the mechanical and electromechanical properties. Concerning ex-PAN fibers the Young's modulus ranges from 200 GPa (HT fibers) to 600 GPa (HM fibers). Ex-pitch fibers have a higher modulus up to 900 GPa. Figure 4 shows the correlation between the Young's modulus and the specific resistivity of different carbon fiber types.

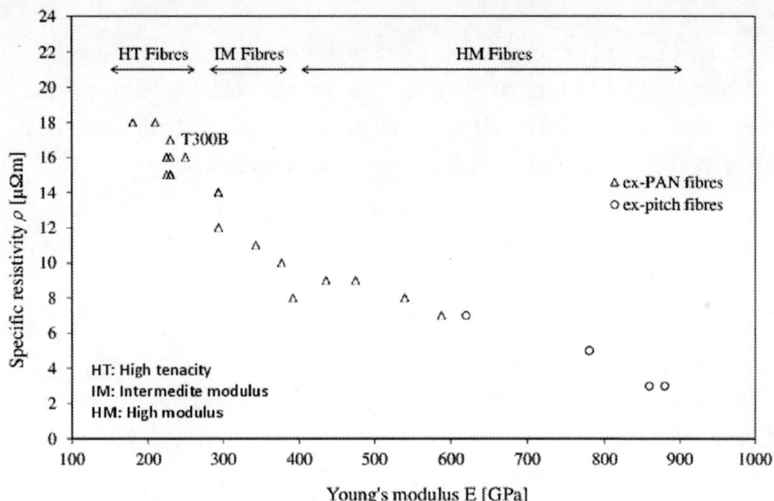


**Figure 4.** Correlation between the specific resistance ρ and the Young's modulus E of different ex-PAN and ex-pitch carbon fibers.

The specific resistivity of HT fibers (high tenacity) is in the range of 15 μΩm up to 18 μΩm. Higher orientated fibers (IM fibers and HM fibers) show a specific resistivity below 14 μΩm.

The specific resistance of carbon fibers is strongly temperature dependent [15]. In consequence, temperature effects have a high influence on the signals of CFSs and must be considered by an appropriate temperature compensation (see section 3).

## Piezoresistivity

The electrical resistance R of a carbon fiber is given by:

$$R = \rho \frac{L}{r^2 \pi} \quad (1)$$

Where ρ. is the specific resistivity, Lthe length and r the radius of the fiber. The total differential of R=f(ρ,L,r) is then yielded by the following Equation (2).

$$\mathrm{d}R = \frac{\partial R}{\partial \rho}\mathrm{d}\rho + \frac{\partial R}{\partial L}\mathrm{d}L + \frac{\partial R}{\partial r}\mathrm{d}r = \frac{L}{r^2\pi}\mathrm{d}\rho + \rho\frac{1}{r^2\pi}\mathrm{d}L - 2\rho\frac{L}{r^3\pi}\mathrm{d}r \quad (2)$$

$$with \quad \varepsilon = \frac{dL}{L} \quad and \quad \nu = -\frac{dr/r}{dL/L} \tag{3}$$

The term dρ/ρ denotes the piezoresistive effect (material effect) and the term ε(1+2v) represents the geometric effects. The strain sensitivity k covers both effects. It should be noted that the strain sensitivity k depends on the effective Poisson ratio v

Therefore, the strain sensitivity $\left(\frac{dR}{R}\right) = k_l\varepsilon_l + k_t\varepsilon_t$ must be defined for a corresponding Poisson ratio.

For some applications of CFSs it can be useful to split the strain sensitivity k=1.71 into the longitudinal strain sensitivity v and the transverse strain sensitivity $k_l$. The relative change in resistance is then given by:

$k_t$ (4)

The strain sensitivity μ of a T300B 1K fiber was determined as v for a corresponding Poisson ratio $v_f$ of 0.28. The sensor fiber exhibits a longitudinal strain sensitivity vp in the range of 1.72 to 1.78 and a transverse strain sensitivity $-\varepsilon_y/\varepsilon_x$ in the range of 0.37 to 0.41. The piezoresistivity of the ex-PAN fiber is linear up to a strain level of approximately 6000 v=0.285m/m [1, 6]. Problems may occur at the metallized fiber endings. In order to avoid any influence of the electrical connections the strain level at the fiber endings should not exceed a level of 2500 μm/m.

This can be realized by using special tabs which exhibit a higher stiffness in the regions of the metallic connections. An example of such load relieving tabs is given in Figure 5. The lay-up of the tabs ensures a four times smaller strain level in the region of the electrical connections compared to the strain level which occurs in the testing area.

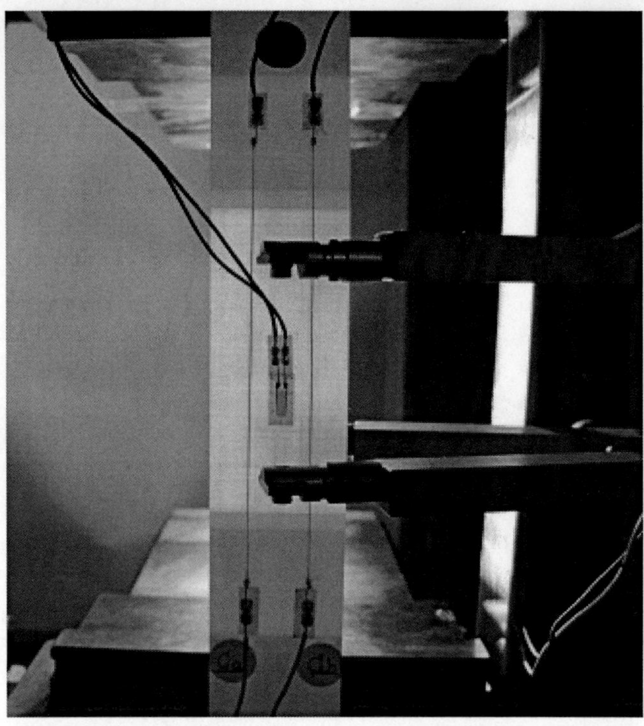

**Figure 5.** CFS patch with load relieving tabs Lay-up in the testing area: [0°] Lay-up in the regions of the metallized fiber endings: $[0°_5]$

Figure 6 shows the excellent linear piezoresistive behavior of the ex-PAN fiber T300B 1K up to a strain level of 6000 µm/m (loading and unloading). Bending tests were performed to investigate the compression behavior of CFSs. These investigations are not completed. First results show that the linearity of the signal depends significant on the carrier material.

Ex-pitch fibers show a nonlinear piezoresistive behavior.

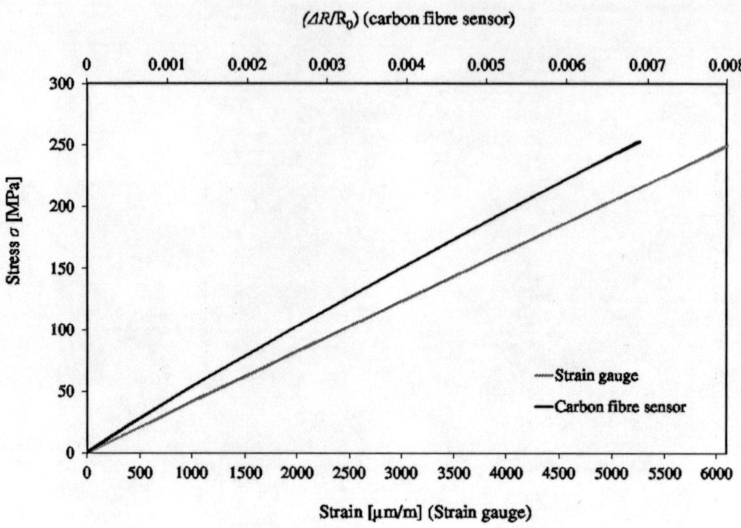


**Figure 6.** Characterization of the linear piezoresistivity up to a strain level of 6000 µm/m (loading and unloading) by means of a CFS patch with load relieving tabs at the endings Sensor fiber: T300B 1K (ex-PAN fiber)Lay-up in the testing area: $[0°_8]$Lay-up of the load relieving tabs: $[\pm 45°, 0°_3]_{sym}$

## THE WHEATSTONE BRIDGE

The change in resistance of a CFS due to an applied strain is usually small. The Wheatstone bridge is an electrical circuit which allows the determination of very small changes in electrical resistance with great accuracy. Furthermore, the Wheatstone bridge minimizes the high influence of temperature changes on the CFS signal. The measurement circuit, illustrated in Figure 7, consists of four resistances, a supply voltage and the output voltage of the bridge.

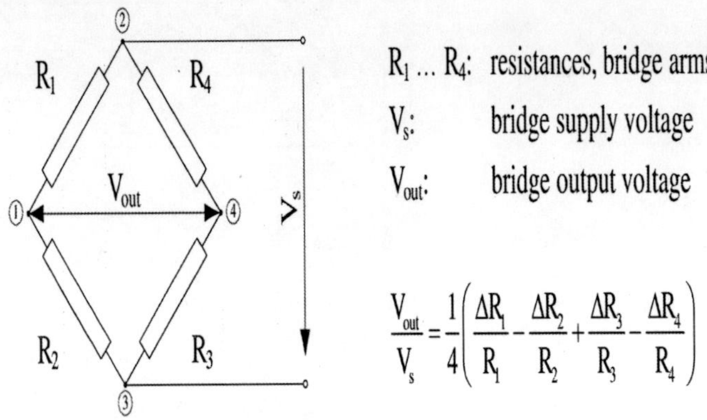


**Figure 7.** General Wheatstone bridge circuit.

The relative change in resistance can be determined by the ratio of output voltage to input voltage $V_{out}/V_s$. There are two configurations of the Wheatstone bridge which are of special importance for the use of CFSs. These configurations of the Wheatstone bridge are known as "half bridge" configuration and "full bridge" configuration.

In the case of a half bridge circuit (Figure 7), the bridge is formed by two CFSs ($R_1$andR$_2$) and two completion resistors ($R_3$andR$_4$). The full bridge configuration (Figure 7) is formed by four CFSs and needs no additional resistors. Both circuits will enable a compensation of temperature effects (thermal dependency of the specific resistance and thermal expansion) if the thermal conditions of the connected CFSs are identical. In the case of a half bridge circuit with one active CFS ($R_1$) the output voltage $V_{out}$ is given by:

$$V_{out} = \frac{V_s}{4}\left(\frac{\Delta R_{1,mech} + \Delta R_{1,therm}}{R_1} - \frac{\Delta R_{2,therm}}{R_2}\right) \quad (5)$$

Equation (5) shows that the thermal effect is compensated by the mechanically unloaded carbon fiber sensor ($R_2$). Detailed information about the principle and use of the Wheatstone bridge can be found in classical textbooks on strain gauge techniques (e.g. [5, 13])

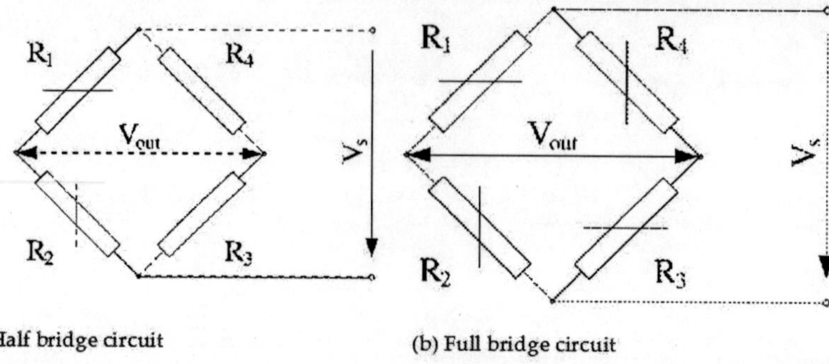


(a) Half bridge circuit (b) Full bridge circuit

**Figure 8.** Half bridge and full bridge configuration of the Wheatstone bridge.

## INTEGRAL STRAIN MEASUREMENT OF THE CARBON FIBER SENSOR

Equation (3) describes the relative change in electrical resistance ($\Delta R/R_0$) of a carbon fiber sensor due to an applied elastic strain ε. It is important to understand that a CFS measures the strain integrally along its whole fiber length. Thus, Equation (3) can be written as:

$$\left(\frac{\Delta R}{R_0}\right) = k\,\frac{1}{L}\int_0^L \varepsilon(x)\,\mathrm{d}x \qquad (6)$$

Equation (6) shows that a CFS measures the displacement between the terminal points of the sensor fiber.

$$\left(\frac{\Delta R}{R_0}\right)\frac{L}{k} = \int_0^L \varepsilon(x)\,\mathrm{d}x = \left[u(x=L) - u(x=0)\right] \qquad (7)$$

This integral strain measurement of CFSs in accordance to Equation (7) can be used to create carbon fiber sensor meshes (CFS meshes). Such a CFS mesh allows the determination of the two dimensional (2D) state of strain and state of deformation of a whole structure or larger areas of a structure.

## CARBON FIBER SENSOR MESHES

The strain and deformation analysis by means of CFS meshes are based on linear or higher order displacement approximations. The displacement functions depend on the used element type such as 3-node triangle elements, 6-node triangle elements or quadrilateral elements. The basic theory of strain analysis with CFS meshes using 3-node triangle elements with a linear displacement approximation is presented below.

A triangular CFS element defined by its vertices 1, 2, 3 and its local coordinate system is given in Figure 9.

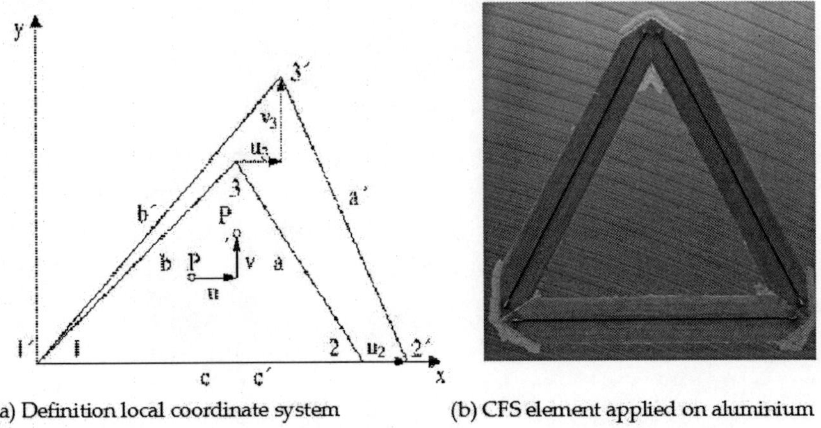

(a) Definition local coordinate system (b) CFS element applied on aluminium

**Figure 9.** Triangle CFS element with 3 nodes.

The displacement u(x,y) and v(x,y) of an inner point P can be determined by the displacements of the vertex in using a displacement function.

Assuming a linear displacement function the displacement u(x,y) and v(x,y) within the element can be calculated in accordance to Equation (8) and Equation (9).

$$u(x,y) = N_1 u_1 + N_2 u_2 + N_3 u_3 \tag{8}$$

$$v(x,y) = N_1 v_1 + N_2 v_2 + N_3 v_3 \tag{9}$$

Hereby $u_i$ and $v_i$ denote the displacements of the vertex 1, 2, 3 in the x- and y-direction while $N_i$ represents the shape functions of the element. The shape functions $N_i$ of a 3-node triangle can be found in classical books on the theory of finite element analysis (e.g. [16]). The engineering strains $\varepsilon_x$, $\varepsilon_y$ and $\gamma_{xy}$ are defined by:

$$\varepsilon_x = \frac{\delta u}{\delta x}; \qquad \varepsilon_y = \frac{\delta v}{\delta y}; \qquad \gamma_{xy} = \frac{\delta u}{\delta y} + \frac{\delta v}{\delta x} \tag{10}$$

Considering the local coordinate system ($u_1$=0, $v_1$=0,$v_2$=0) the strains within the triangle can be calculated by:

$$\varepsilon_x = \frac{u_2}{x_2}; \qquad \varepsilon_y = \frac{v_3}{y_3}; \qquad \gamma_{xy} = \frac{x_2 u_3 - x_3 u_2}{x_2 y_3} \tag{11}$$

In consequence of the linear displacement approximation the strains are independent of the coordinates (x and y) and thus constant within the element. The unknown displacementsu2, u3and v3 of the vertex 2 and 3 can be determined by the signals of the carbon fiber sensors.

In order to verify this linear approach an experimental investigation of a CFS mesh applied on a 1000 mm x 1000 mm x 5 mm PMMA plate was performed. The simply supported plate was loaded with a single static force at the center. In addition to the experiment a finite element analysis (FEA) was performed. In [9] the results of this experiment and of the corresponding finite element simulations are presented in detail [5].

Figure 10 shows the PMMA plate and the applied CFS mesh.

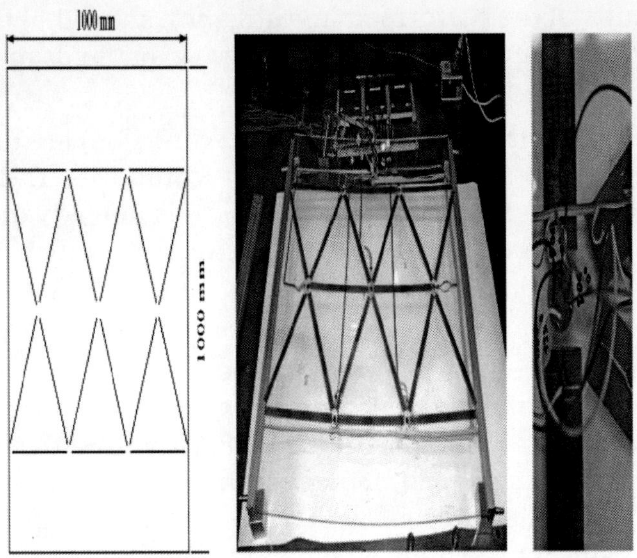


**Figure 10.** Carbon fiber sensor mesh applied on a 1 m x 1 m PMMA plate. Each sensor has a length of 300mm. [9].

Figure 11 shows the determined strain εx for each element of the mesh. There was a good correlation between the measured and the calculated strain levels. The accuracy was in the range of±5%.

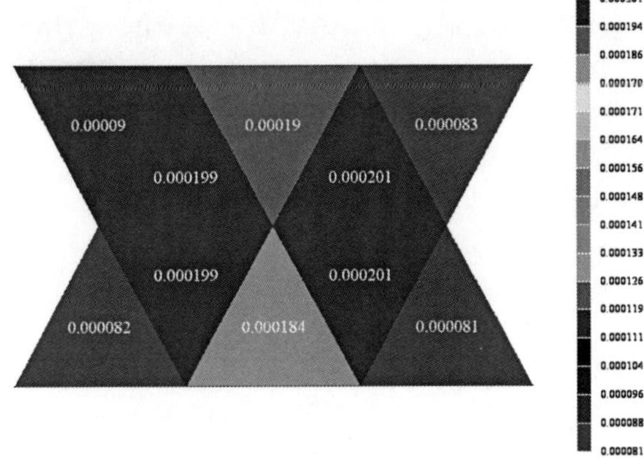


**Figure 11.** Strain $\varepsilon_x$ measured by the carbon fiber sensor mesh.

The results of the performed investigation show that CFS meshes are a reliable instrument to determine the strain fields and principle strains of lightweight structures.

The principle strains (strain level and direction) are of particular interest in case of structures made of composite materials. For example, tailored fiber placement (TFP) is an advanced textile manufacturing process for CFRP structures in which the carbon fiber rovings are placed in accordance to the direction of principal stresses.

The finite element simulation is a powerful tool to analyze the stress fields and principle directions of lightweight structures. At the design and optimization processes the FEA is almost the only way to evaluate the structural load. However, there is a lack of techniques to review the results of the FEA. CFS meshes offer a high potential to verify the results of the finite element analysis.

## MICRO CRACK DETECTION

A major failure mode of multidirectional reinforced laminates is transverse matrix cracking. Matrix cracks will reduce the effective stiffness of the laminate and will result in local stress concentrations at the crack tip. Furthermore, interlaminar crack growth and local delamination can occur. Due to its integral strain measurement method the CFS has a high potential to detect matrix crack initiation and monitor crack growth. Figure 12 shows a thin GFRP laminate (Lay-up: [$90^{\circ}_{2}$, 0°, $90^{\circ}_{2}$]) which has two embedded CFSs. Matrix cracks along the CFS will affect the sensor signal which will give a clear indication of the crack density. A study was performed to investigate the influence of cracks on the sensor signal [10].

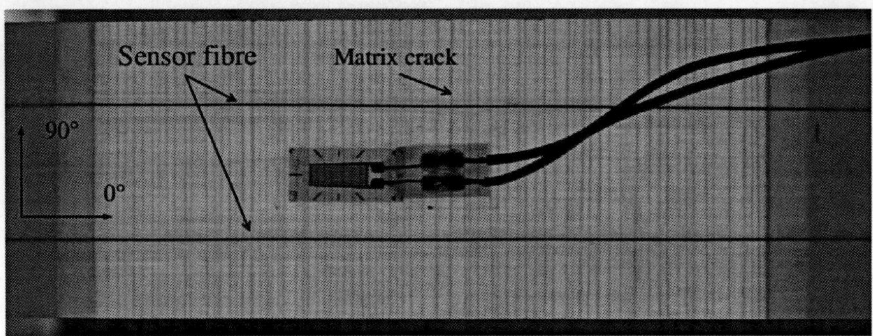


**Figure 12.** Multidirectional reinforced GFRP laminate with two embedded CFSs. Transverse matrix cracking will affect the sensor signal. Lay-up: [$90°_2$, 0°, $90°_2$]

The study was performed on the GFRP laminate[0°,$90°_5$,0°,$90°_5$,0°]. Three CFSs were embedded in the mid-plane 0°-layer. At a strain level higher than 3000 μm/m matrix cracks appeared in the 90°-layers. The following techniques were applied to characterize the influence of damages on the sensor signal:

- Acoustic emission analysis in combination with pattern recognition technique
- Microscopy and micrographs
- Analytical calculations
- Finite element analysis

Figure 13 shows the correlation between the crack density, the CFS signals, the acoustic emission energy (AE-energy), the strain level measured by a strain gauge and the according stress level. It can be seen that the matrix crack initiation causes first AE-signals and a change of the slope of the CFS signals.

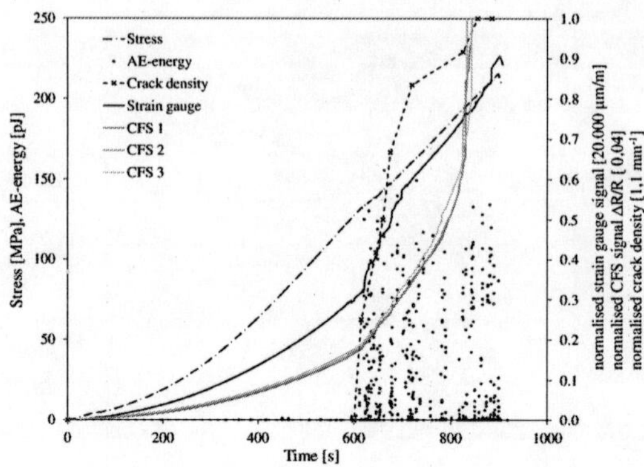


**Figure 13.** Acoustic emission and CFS signals measured on an GFRP laminate under uniaxial tensile load [10] Lay-up: [0°,90°$_5$,0°,90°$_5$,0°].

Furthermore, a good correlation between the sensor signal ($\Delta R/R_0$) and the crack density can be observed. After having reached a crack density of 0.8 mm-1 the signals of the embedded carbon fiber sensors increase disproportionately. The analytical approach of Garret and Bailey [4, 11, 12] and a FEA were applied to calculate the reduction of the stiffness of the laminate. Müller showed that the global stiffness loss of the laminate due to the matrix cracking can be measured by means of the CFS. However, at high strain levels (>70% of $\varepsilon_{ultimate}$) there is a strong influence of high local stresses at the crack tip on the CFS signal. These stress concentrations may result in filament breakage and in extremely high signal levels.

Based on this result CFSs can be used for damage monitoring and for the prediction of the lifetime of damaged structures if the damage level can be characterized by stiffness loss. Figure 14 shows the cyclic loading of a laminate to measure the stiffness loss and to determine Ladeveze's material parameters [8]. Based on this calibration the lifetime of a structure, e.g. pressure vessel, can be predicted.

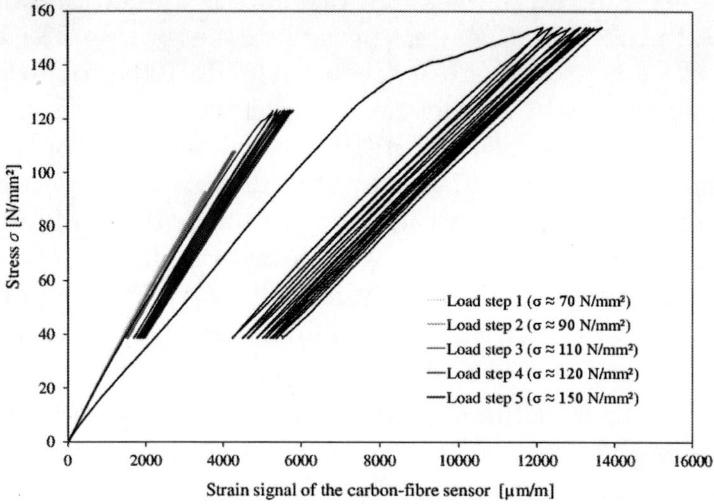


**Figure 14.** Damage measurement and determination of damage variables like the energy release rate Material: GFRP, EG/913 Lay-up: [0°,90°$_5$,0°,90°$_5$,0°].

## IMPACT AND DELAMINATION DETECTION

An impact loading can cause small damages inside the composite material which may not be found by visible inspection. One major concern is delamination damages or the dis-bonding of interfaces. These damages result in sub laminates having lower buckling resistance and compression strength. Although a small delamination-damage does not necessarily constitute failure, the damaged area may undergo a time-dependent growth and may attend a critical size.

The principles for achieving damage tolerant primary composite structures were established by the aircraft companies [14]. Maintenance intervals and inspection plans are determined in such a way that readily detectable damages will be repaired before damage growth can affect the fatigue strength of the structure. The influence of undetectable damages is covered by the so called barely visible impact damage (BVID) which defines the damage that establishes the strength values to be used in analysis to demonstrate compliance with the load requirements. One method to determine the influence of the BVID on the mechanical performance is the compression after impact test procedure (CAI-test procedure, i.e. Boeing BSS 7260). A 4 mm thick quasi-isotropic test specimen (150 x 100 mm) is damaged by a dropped weight impact testing machine. An impact level of about 3 J/mm will cause the BVID. This means that only a small remaining indentation is visible on the surface of the specimen, but delamination may be found inside. The strength of the damaged specimen is reduced by 10 to 20% compared to the undamaged material. This shows that in many cases the exploitation of material performance is limited, since the skin thicknesses of a composite structure are designed to absorb an impact.

For aircraft structures health monitoring systems have an extremely high potential to improve the efficiency of composite structures. Based on a monitoring system the material design values can be increased and the inspection intervals can be enlarged. Both aspects can result in a weight reduction of the structure up to 10%. A preliminary study was performed to investigate the use of CFSs as a sensor system to detect the BVID. The CAI specimen was used to integrate a rectangular CFS mesh (Figure 15 and Figure 16).

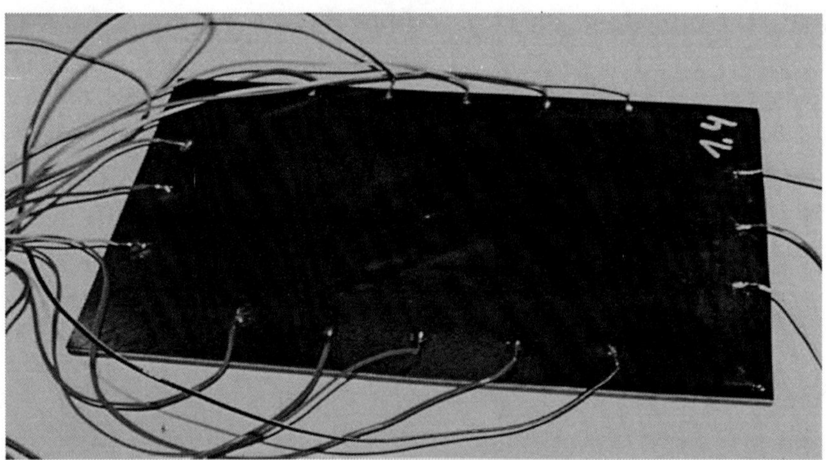


**Figure 15.** Compression after impact (CAI)-specimen with integrated CFS mesh to detect damages below the barely visible impact damage level (BVID).

The distance between the CFSs varied from 30 to 100 mm. Three methods were investigated to detect the impact damage:

- Online measurements of the resistivity during the impact test
- Offline measurements, comparison of the resistivity before and after the impact
- Active thermography, CFSs used as heating element

It has been shown that all three procedures are suitable to detect impact damages. A distance of 50 mm between the CFSs is necessary to detect even small damages below the BVID. In a second step the use of CFSs will be investigated to detect debonding of skin and stringer. A health monitoring system for a complex aircraft structure will be based on several technologies like ultrasonic inspection sensors and strain sensors. The CFS technology will complement the established sensors due to its specific and simple integral measurement principle.

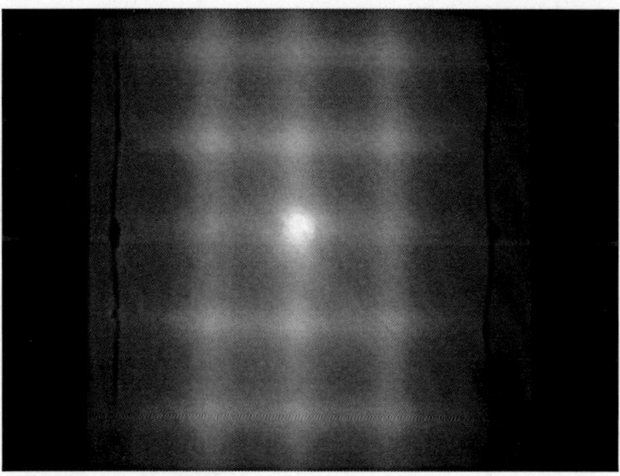

**Figure 16.** Compression after impact (CAI)-specimen with integrated CFS mesh. The CFSs are used as heating element for active thermography. A small delamination damage is visible (BVID).

## APPLICATION

CFSs offer a high potential to be used as a sensor element for composite materials for stress analysis, damage detection and the monitoring of manufacturing processes. Two industrial applications have been selected to demonstrate this.

## Tabletop of a CT-Scanner

Carbon fiber reinforced plastics (CFRP) are used for tabletops of computer tomography (CT) scanners, since CFRP fulfills the X-Ray transparency which is necessary to get the picture quality sufficient for medical diagnosis.

CFSs can be embedded in the tabletop of a CT scanner to measure its deflection. Based on the measured deflection the CT images can be readjusted to result in an improved medical attendance [7]. Figure 17 shows the tabletop of a CT scanner with ten u-shaped CFSs applied. For this application u-shaped CFSs are used to avoid that any metal wiring is part of the scan plane for all operation positions.

**Figure 17.** CT tabletop with u-shaped CFSs to measure the deflection. The metallic wiring is attached to the clamping support.

The lengths of the u-shaped CFSs vary from 250 to 1250 mm. The determination of the beam deflection is based on the integral strain measurement of CFSs (Equation (6)). Considering small deformations the relation between the elastic strain of the outer fiber

$\varepsilon\hat{}$ and the beam deflection v is given by:

$$v = \iint v'' dx dx = \iint \frac{\hat{\varepsilon}}{e_y} dx dx \quad (12)$$

where $e_y$ denotes the distance from the neutral axis. Assuming a cantilever beam (see Figure 18) the slope v′ of the beam can be determined directly by the signals of the CFSs. Assuming that the CFS starts at the clamping support (x=0) the slope v′ at the end of the applied CFS ($x=l_i$) becomes:

$$v'(x = l_i) = \frac{l_i}{key} \left(\frac{\Delta R}{R}\right)_i \quad (13)$$

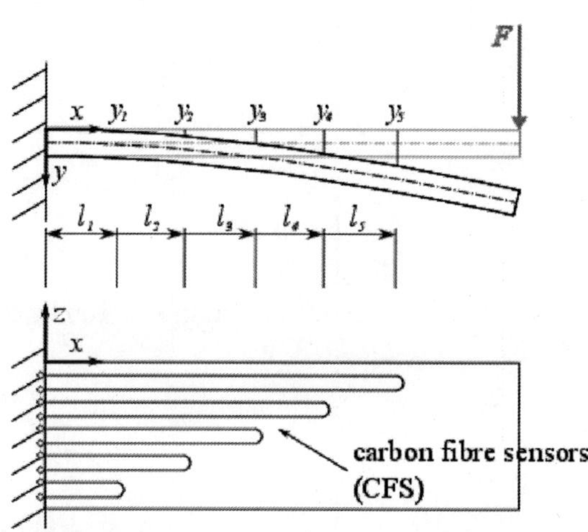


**Figure 18.** Side and top view of a cantilever beam with five u-shaped CFS.

The deflection y of the beam can be calculated by numerical integration.

$$v(x = l_i) = \sum_{n=1}^{i} \left[ v'(x_n) - \frac{v'(x_n) - v'(x_{n-1})}{2} \right] (l_n - l_{n-1}) \quad (14)$$

The index *i* denotes the number of CFSs applied. The quality of the approximation in accordance to Equation (14) depends on the complexity of the loading, the number of applied CFSs and the used sensor configuration.

In the case of the CT table (Figure 17) the deflection of the table can be determined with an accuracy of ±0.3 mm for different operation positions. Figure 19 shows a typical measurement. The operating position of the CT tabletop varies stepwise from 0 mm to 2000 mm and back to 0 mm. The weight of the used dummy is 100 kg and electrical half bridges are used for temperature compensation.

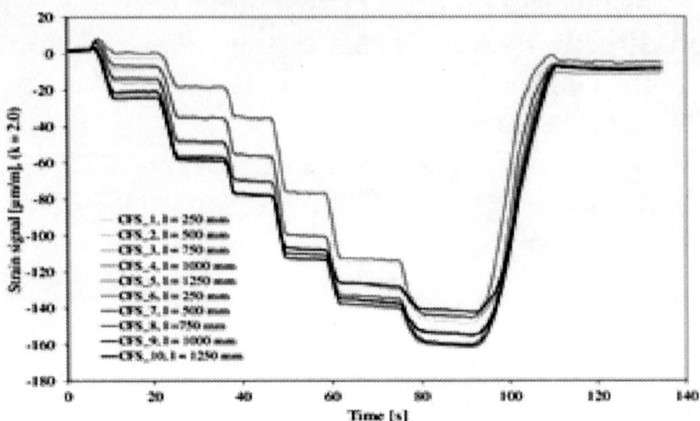


**Figure 19.** CFS signals of the CT table for a weight of 100 kg and different operation positions, ranging from 0 to 2000 mm.

## Pressure Vessels

For thin walled assumptions the longitudinal stress $\sigma_l$ and hoop stress $\sigma_r$ in the cylindrical portion of a pressure vessel, away from the ends, are given by:

$$\sigma_l = \frac{pr}{2t} \qquad (15)$$

$$\sigma_r = \frac{pr}{t} \qquad (16)$$

where $t$ is the thickness and $r$ is the radius of the vessel. The relation $\sigma r/\sigma l$ shows that the efficiency of pressure vessels can be increased if hoop wrapped vessels are used. A metal cylinder is reinforced by carbon fibers having a radial orientation (type II vessels). The strength of the vessel is increased remarkably.

There are two aspects for the use of CFSs for pressure vessels which can be easily integrated by means of the winding process:

- Determination of the pressure level of the vessel
- Monitoring of degradation processes due to fatigue or overloading of the radial fiber reinforcement

For such an application a sensor patch with four CFSs connected in full-bridge configuration is particularly suitable, since the full-bridge circuit minimizes the influences of thermal effects and shows an improved long term stability of the signal. Figure 20 shows such a hoop wrapped vessel with an embedded CFS patch.

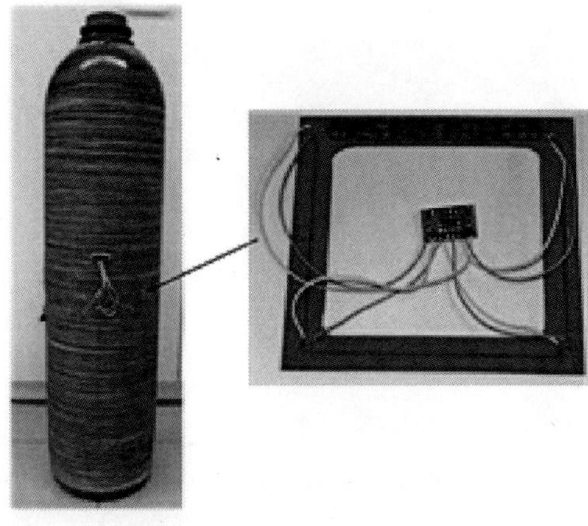

**Figure 20.** Hoop wrapped pressure vessel (type II vessel) with embedded CFSs.

Experimental studies showed that the pressure of the vessel can be determined by using a CFS patch with a resolution of about 1 bar. A representative measurement of the performed test is shown in Figure 21. The pressure load of the vessel was increased stepwise up to a maximum pressure level of 100 bar. The subsequent pressure relief was performed in the same manner.

Micro cracks in the CFRP layers of the pressure vessel may occur as a result of mechanical or thermal overloading. The failure of the matrix causes a loss of stiffness of the CFRP layers and reduces the global stiffness of the vessel. Depending on the crack density and the crack growth the lifetime of the structure will decrease. Based on the damage master curve of the structure, an estimation of the remaining lifetime can be performed (see Figure 14).

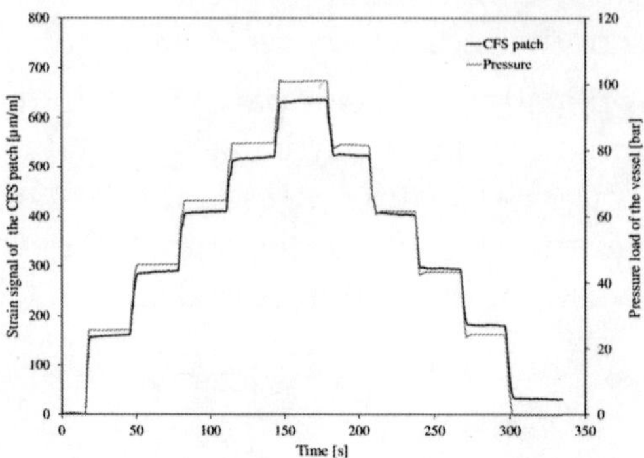


**Figure 21.** Strain signals of a CFS patch applied near the surface of a type II pressure vessel subjected to a pressure test.

# CONCLUSION

There are three main aspects which made carbon fiber sensors (CFSs) very interesting to be used for composite materials:

- Material-conformity
- Linear piezoresistivity up to high strain levels
- Integral strain measurement method

CFSs based on a T300B 1K ex-Pan fiber exhibit an excellent linear piezoresistivity up to a strain level of 6000 µm/m. The according strain sensitivity was determined as k=1.71 (related to a Possion ratio of $\nu$=0.28). The longitudinal strain sensitivity $k_l$ of the CFS is in the range of 1.72 - 1.78. Transverse to the fiber direction CFSs exhibit a transverse strain sensitivity $k_t$ of approximately 0.4. This significant transverse strain sensitivity must be considered in praxis. Tension load was applied for all tests, the characterization of the compression behavior is under examination.

A disadvantage of CFSs is the high influence of temperature on the signal which will be an aspect for future research. At the moment a long term stability of.±2 µm/m (T=*const.*) can be achieved for a half or full bridge circuit.

CFSs can be used for strain analysis, damage monitoring and

the control of manufacturing processes. Concerning strain and stress analysis an approach for CFS meshes was developed based on triangular elements with linear displacement approximation. A good correlation was found between the measurement and the finite element calculation. The use of CFS meshes can be a new approach to complement finite element analysis from the experimental side.

Concerning monitoring aspects the influence of material damages on the sensor signal were studied. It has been demonstrated that based on the integral strain measurement method the CFS is an excellent sensor to detect delaminations and matrix cracks in multidirectional reinforced laminates. Therefore, the CFS offers unique features for fracture mechanics: Measurement of strain levels and detection of matrix cracks. By means of two examples the CFS technologies could be demonstrated successfully:

- Determination of the deflection of a tabletop of a CT-Scanner
- Determination of the strain level and the crack density of a pressure vessel

Strain measurement and matrix crack detection are in the focus of safe and damage tolerant composite structures making the CFS technology a complement of established sensors.

## Notes

[1] Piezoresitivity describes the change in electrical resistance of a conductor due to an applied strain.

[2] For example, the epoxy resin EP 310 S can be burned off or removed using acid.

[3] The effective Poisson ratio k depends on the Poisson ratio of the sensor fiberk, the Poisson ratio of the sensor patch kl and the strain ratio kt of the structure.

[4] Conventional strain gauges exhibits the same behaviour. The strain sensitivity of strain gauges is usually defined for a corresponding Poisson ratio k (steel).

[5] A quadratic displacement approach for a 3-node triangle is also presented and verified in [9]. However, the quadratic displacement approach of a 3-node triangle requires additional strain gauges at the nodes to determine the 12 unknown

coefficients of the shape function.

## REFERENCE


1. Bibsonomy CiteULike Reddit LinkedIn StumbleUpon Mail to a Friend
2. C. Christner, A. Horoschenkoff, H. Rapp, 2012Longitudinal and transverse strain sensitivity of embedded carbon-fiber sensors, Journal of Composite Materials Online First: DOI:
3. D. Chung, 1994Carbon Fiber Composites, Butterworth-Heinemann.
4. M. Dresselhaus, G. Dresselhaus, K. Sugihara, I. Spain, H. Goldberg, 1988Graphite Fibers and Filaments, Springer.
5. K. Garrett, J. Bailey, 1977Multiple transverse fracture in 90° cross-ply laminates of a glass fiber-reinforced polyester, Journal of Material Science 12
6. K. Hoffmann, 1987An introduction to measurements using strain gages, Hottinger Baldwin Messtechnik.
7. A. Horoschenkoff, T. Müller, A. Kröll, 2009On the charaterization of the piezoresistivity of embedded carbon fibers, 17th International Conference on Composite Materials, Edinburgh.
8. A. Horoschenkoff, T. Müller, C. Strössner, K. Farmbauer, 2011Use of carbon-fiber sensors to determine the deflection of composite-beams, 18th International Conference on Composite Materials, International Conference on Composite Materials, Jeju.
9. P. Ladeveze, E. L. Dantec, 1992Damage modelling of the elementary ply for laminated composites, Composite Science and Technology 43
10. T. Matzies, C. Christner, T. Müller, A. Horoschenkoff, H. Rapp, 2011Carbon-fiber sensor meshes: Simulation and experiment, 21st International Workshop on Computational Mechanics of Materials, Limerick.
11. T. Müller, A. Horoschenkoff, H. Rapp, M. Sause, S. Horn, 2010Einfluss von zwischenfaserbrüchen in 0/90laminaten auf die elektrische widerstandsänderung von eingebetteten carbonfasern, 59th Deutscher Luft- und Raumfahrkongress.
12. J. Nairn, 1989The strain energy relase rate of composite microcracking: A variational approach, Journal of Composite Materials 23
13. J. Nairn, S. Hu, 1994Matrix microcracking, Damage Mechanics of Composite Materials 9
14. C. Perry, H. Lissner, 1955The Strain Gauge Primer, Mc Gram Hill.
15. H. Razi, S. Ward, 1996Principles for achieving damage tolerant primary

composite aircraft structures, 11th DoD/FAA/NASA Conference on Fibrous Composites in Structural Design.

16. I. Spain, K. , V. , H. Goldberg, I. Kalnin, 1982Unusual electrical resistivity behavior of carbon fibers, Solid State Communications 817-819

17. O. Zienkiewicz, R. Taylor, 2000Finite Element Method 1The Basis, Elsevier.

# Chapter 11

# COMPOSITE MATERIAL AND OPTICAL FIBERS

Antonio C. de Oliveira [1] and Ligia S. de Oliveira[1]

[1] Laboratório Nacional de Astrofísica, Ministério da Ciência Tecnologia e Inovação, Brazil

## INTRODUCTION

For ease of handling and polishing, optical fibers are generally mounted in some form of rigid structure. A schematic of a typical fiber termination used in astronomical instruments (typical of multiple-fiber spectrographs) is show in Fig. 1. The fiber is first placed within a flexible tube (polyimide or similar), often referred to as the strain relief tube. The fiber and tube are placed within a rigid ferrule. Adhesive is applied to the fiber and tubing to fix them in place. The ferrule can then be easily manipulated for polishing, or mounting within an instrument, without risk of damage to the fiber. The role of the strain relief tube is to prevent stresses occurring at the point where the fiber enters the ferrule given bending at this point may lead to breakages. For coupling an array of optical fibers to a microlens array, such as in IFU (Integral Field Unity), a brass plate with an array of drilled microholes may be used. Optical fibers positioned in an array of accurately drilled holes, as illustrated schematically in Fig. 2.

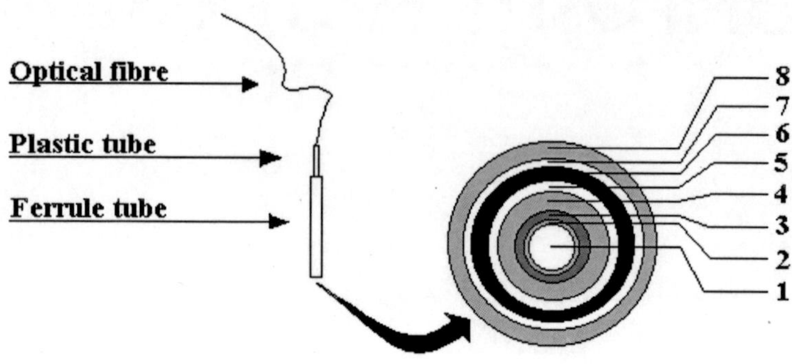


**Figure 1.**Schematic diagram of a single fiber mounting assembly: 1, core; 2, cladding; 3, polyamide buffer; 4, acrylate buffer; 5, epoxy; 6, plastic tube; 7, epoxy; 8, ferrule steel tube.

This system, called microholes array, contains a grid of holes spaced by the pitch of the microlens array. The holes are machined using custom made drills with two different diameters. This produces a stepped hole, with the smaller diameter hole used for fiber positioning while the larger hole is used to accommodate a ferrule. The small holes are approximately 10 µm larger than the fiber diameter to allow sufficient space for a glue to penetrate. Using a stepped hole also allows a greater depth of material to be machined than by using a small drill alone. This permits a thicker, hence more robust, piece of material to be used. A support plate is also used, positioned above the fiber-positioning array with spacers, to maintain accurate angular alignment of the ferrules with respect to the microlens optical axis. Each ferrule contains a polyimide strain relief tube to prevent mechanical stress, occurring at the point where the fiber enters the ferrule. To secure the fibers, ferrules, and polyamide tubes in place the whole input assembly is immersed in a container of EPOTEK 301-2 adhesive. This epoxy is a natural choice due to its excellent wicking properties and low shrinkage upon curing. After curing, which takes approximately three days at room temperature; any excess glue is removed prior to optical polishing.

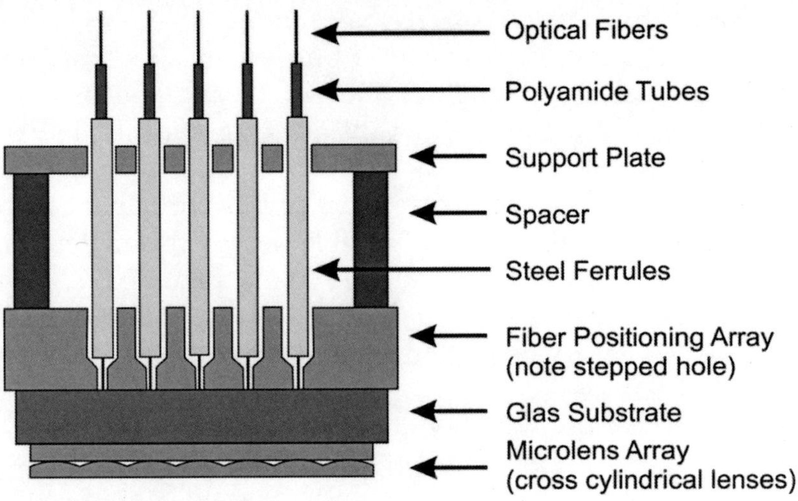


**Figure 2.** Schematic micro lens array and fiber positioning array used to constructed IFUs.

The problem here may be an increase of FRD (Focal Ratio Degradation) caused by contraction of the metal ferrule or brass plate at low temperature causing stress on the fibers and consequent loss of throughput. Astronomic instruments like that in general work in environments with significant thermal gradients, a common characteristic of ground-based observatories. An interesting alternative to the conventional steel ferrule may be a quartz tube. Quartz material has no problems of contraction in the temperature gradients experienced in that places, -10 °C to 20 °C, but is very expensive and difficult to obtain. The ideal condition requires a material with elasticity controlled so as not to cause stress or shift the positioning of optical fiber under temperature gradients. For just such purposes, we have developed a special composite formed from a mixture of EPO-TEK 301-2 and some refractory material oxide in nano-particle form, cured and submitted to a customized thermal treatment. To avoid bubbles and points of stress, this mixture is prepared in a separate receptacle inside a vacuum chamber. The resulting material is more resistant and harder than EPO-TEK 301-

2 and is found to be well suited to the fabrication of optical fiber arrays. An important secondary characteristic is the ease with which it can be polished. This feature is a result of the micro particles, which keep the polished surface very homogeneous during the final polishing procedure. The resulting composite combines the beneficial characteristics of both the epoxy and the oxide; main factor its coefficient of thermal expansion is significantly lower than simple solidified epoxy; the exact value depending on the relative concentrations. While the characteristics of this particular composite are still under study, it is clearly possible deploy this material in the construction of devices for several fiber instruments.

## New Materials to Support Optical Fibers

Similar microholes arrays and the support plates of the Fig. 2, used to construct Eucalyptus IFU, were made with toolmakers brass (de Oliveira et al., 2002). The problem here is that differential expansion between the metal array and the glass microlens substrate may lead to the bond between them failing at low temperatures. Although the coefficient of thermal expansion of epoxy is much greater than those of steel, brass or glass, the elasticity of the epoxy accommodates the dimensional changes without breakage: however, this can introduce a small amount of stress build-up.

It is well known that mechanical deformation causes focal ratio degradation (FRD) by the formation of microbends in the fiber (Clayton 1989). FRD is a non-conservation of *étendue* such that the focal ratio is broadened by propagation in the fiber. When mounting the fiber, the appropriate epoxy and tubing should be selected, and general care must be taken to minimize mechanical stress and avoid additional FRD (de Oliveira et al., 2005). This is straightforward at room temperature, but greater care must be taken in the choice of materials for use at low temperatures. When the fiber assembly is cooled to temperatures around -10 °C, the epoxy, tubing and ferrule will all shrink differentially, and this may cause the level of FRD to increase. There is other problem when the fiber assembly is warmed to around 20 °C and cooled to around -10 °C. The UV epoxy, currently used to cement the metal and the glass may be damaged if the system will be submitted several times at big changes of temperature. In this case the system may detach in places due the thermal gradient.

Microholes array and support plate made with solidified EPOTEK represent a first step to change the metallic base for a polymeric base. The choice of the EPOTEK 301-2 is appropriate due to its excellent wicking properties and low shrinkage upon curing. This properties are very good to use in optical fibers system, so that, if it is possible to get plates adequate to machine with this epoxy we can get total compatible in the construction of the system.

## Epoxy solidified

The epoxy EPO-TEK 301-2 has low viscosity and requires a container to constrain its flow until it is solidified. Generally aluminium has been used to make these containers but it is possible to use brass, plastic or acrylic. The complete curing process takes approximately three days at room temperature and when it is dry, it is transparent to visible light. To avoid bubbles and points of stress, the epoxy is prepared in a separate container inside a vacuum chamber. The correct amount is allowed to set in the container, which is placed inside a dry environment. Once cured, some thermal treatment may be necessary. Several kinds of blocks, cylinders and plates can be made to test the polishing qualities of these test pieces. The results are very encouraging with the hardness similar to acrylic resin. The Fig. 3 and 4 shows steps to obtain samples machined to manufacture blocks of epoxy solidified.

**Figure 3.**EPO-TEK 301-2 solidified, during the machining procedure.

Machining quality is important as burrs inside the microholes may prevent the fibers from being threaded into the holes, or cause stress, or breakage, of the fibers. The quality of such microhole arrays inspected visually using a microscope, give very encouraging results displaying a minimum of burring; scarf, remaining in the holes, may be readily removed by cleaning in an ultrasonic bath.

There are several advantages to the use of solidified epoxy as compared to brass, for example, in the fabrication and use of fiber support devices. Ease of machining and compatibility with other epoxies used to attach glass or silica, may be the most important of these advantages. Although the coefficient of thermal expansion of epoxy is much greater than that of steel or glass, Tab. 1, its elasticity accommodate thermally induced dimensional changes without breakage. This also avoids excessive stress associated with increases in FRD but, in principle, could be deleterious in compromising the critical positioning stability of optical fibers as the temperature varies.

**Figure 4.**Plate of EPO-TEK 301-2 solidified and machined.

**Table 1.**Coefficient of Thermal Expansion of some materials

| *Material* | α *CTE at 20 °C* | *Units* |
|---|---|---|
| *Brass* | *19* | $10^{-6}$ / °C |
| *Carbon Steel* | *10.8* | $10^{-6}$ / °C |
| *In ox Steel* | *17.3* | $10^{-6}$ / °C |
| *Quartz* | *0.59* | $10^{-6}$ / °C |
| *EPO-TEK 301-2* | *55 at 61* | $10^{-6}$ / °C |

Machining quality is important as burrs inside the microholes could prevent the fibers from entering the hole, cause stress, or breakage, of the optical fibers. The quality of the microholes array may be inspected visually using a binocular microscope and the results are often very satisfactory with minimal burring present. Swarf present in the holes is readily removed by cleaning in an ultrasonic bath. The Fig. 5 shows a sample of epoxy solidified with a microholes array.

**Figure 5.**Photograph of the microholes array in a sample of the EPO-TEK 301-2 solidified.

After machined, this sample has a diameter of 48 mm and a thickness of 3 mm. This microholes array is matrix 30x30 holes spaced on a 1.0 mm pitch. The holes were machined using custom made drills with diameters of 0.60 mm and 0.21 mm. This produces a stepped hole, with the smaller diameter hole used for fiber positioning while the larger hole is used to accommodate a ferrule. The small holes are approximately 10 μm larger than the fiber diameter to allow sufficient space for glue penetrates. The machining error in the position of the small holes was measured to be approximately 2 μm.

## Experimental Stress Analyses

It is possible to use a very simple experiment of Photo elasticity method to evaluate the static stress in the plates of epoxy solidified. Classical two-dimensional photo elasticity is an optical experimental technique for determining stress fields in solids bodies. For a given analysis, polarized light is passed through a transparent sample or the body in question, and stress-induced or static stress changes in the light result in an interference-like pattern, which may be analysed to determine the principal stresses at each point within the body.

The basis for photo elastic measurement is a phenomenon of double refraction (also called artificial birefringence). There are two situations that may cause this phenomenon. Certain plastics exhibit the first situation when the sample of this type is subjected to an applied load, the resulting stress / strain field causes the molecules within the transparent material to have a preferred alignment. The second situation is exhibited by certain epoxies after dried and in this case the stress may be called static stress. In both situations the light wave vibrations have two preferred directions within the material and a wave of linearly polarised light entering the field is split into two waves which are linearly polarised at right angles to each other and which propagate with different velocities. That is, two rays travel along each an original line of propagation, and their electric vectors are mutually perpendicular. In fact, each vibration is collinear with one of the principal stress directions. (See Fig. 6.) Also, since the two waves travel at different velocities, a phase difference develops between then and by using certain optical elements.

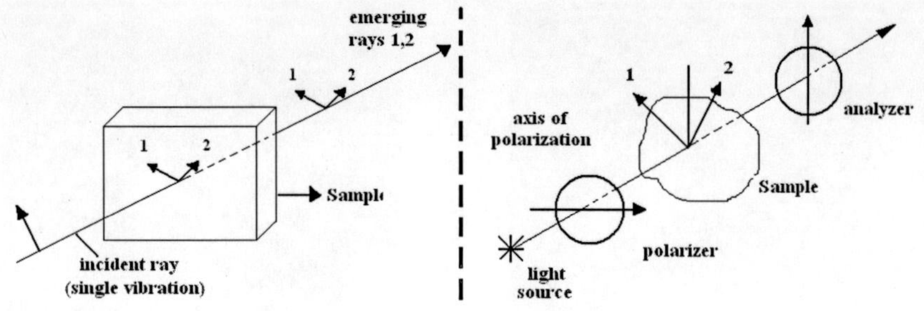


**Figure 6.**Left: propagation of light through photo elastic models. Right: plane polariscope used to get information of the stress inside of the sample by analyse of the artificial birefringence.

A very simple system to get qualitative images of the samples may be adapted with a transmission Polariscope, as shown in Fig. 6. The system uses a CCD camera to take the images. Stresses within solidified EPO-TEK 301-2 may be investigated with the use of the photo elasticity method whereby stress-induced birefringence is measured using polarized light. Static stress can be recorded as an interference pattern, which may be analysed to determine the principal stresses at each point within the material. Indeed, significant stress induced birefringence is detected in such samples, as is shown in Fig.7 side left.

This stress can be alleviated through thermal shock induced by warming the sample to 80 °C for 30 min. Fig. 7 side right demonstrates the reduction in stress as the material is returned to room temperature. Of course, such experiments are only viable for transparent materials but they do give a warning that care must be taken in analysing the effects of thermally induced stress through measurement of fiber displacement in arrangements and FRD stress-induced.

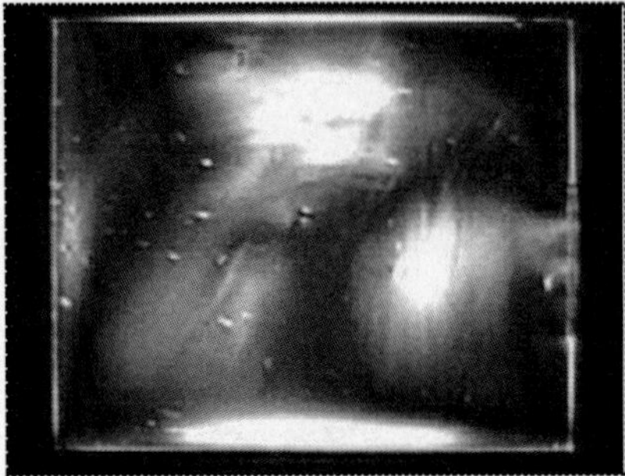

**Figure 7.**Left: sample before thermal shock. Right: same sample after thermal shock.

## Composite

It is possible create composites using a mix of epoxy and several types of oxides in micro or nano-particle form. To avoid stress points and heterogeneous regions, the composite needs to be prepared using mixers of high speed. Ultrasonic chamber can be useful to ensure more uniformity to the mixture. Before the cure, this composite requires be subjected to a vacuum of $10^{-3}$ Torr to reduce bubbles inside of material. The mixture of EPO-TEK 301-2 with refractory material oxide in nano-powder, cured and submitted to a thermal treatment around 400 °C, produce a very interesting option instead simple epoxy solidified. The resulting material is more resistant and harder than EPO-TEK 301-2 and is found to be well suited to the fabrication of optical fiber arrays. Several different refractory material oxides in nano-powder may be used to produce different characteristics in this type of composite. So it is possible combine Zirconium oxide, Barium oxide, Silica oxide, Cerium oxide and others, Fig. 8, to obtain a material optimized to specific applications. The solidified mixture combines the beneficial characteristics of both the epoxy and the oxide; main factor its coefficient of thermal expansion is significantly lower than simple solidified epoxy; the exact value depending on the relative concentrations.

There are two important factors that consolidate the structure of this composite. The first is the process of cure of the liquid mixture. The second is the process of heating of the solid material obtained after the cure. The chemical reactions during the first process are limited by the time to reach the complete cure of the epoxy. Anyway, chemical analysis showed no evidence of endothermic or exothermic chemical reactions between oxides and epoxy. In fact, the materials involved in the mixture appear quite neutral. However, the heating procedure in temperatures around 400 C with slow cooling during 24 hours induces slight shrinkage on the material. Although the study still lacks depth, it is fairly simple to conclude that the structure undergoes some type of molecular rearrangement with some material loss and subsequent compaction. In fact this process carbonizes the external side of the solid material. To avoid total and destructive carbonization, both, the heating and cooling is done with the composite inside a steel container with refractory sand. After this procedure, the external part carbonized can be removed by machining process leaving the sample completely clean. The material thus obtained proves to be quite stable and resistant even though it has some degree of slow oxidation on its surface. This oxidation is evident from the slight colour change after a few weeks of exposure and manipulation, but still remains a high physical stability.

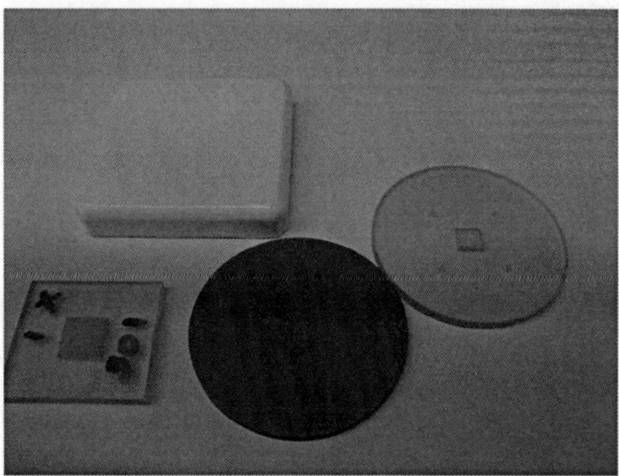

**Figure 8.** Samples of different composites at the centre and epoxy solidified at the borders.

This composite has two physical characteristics very interesting for the construction of optical fibers holders. The first feature is its ability to sustain their polishing, with minimum quantities of abrasives during this procedure. In other words, when the composite is subjected to a polishing of high performance, the detachment of the refractory oxide nanoparticles reinforces gently the polishing process and increasing the efficiency of this procedure. The surface roughness measured in several samples, after high performance polishing was about 0.01 microns. Furthermore, the time for obtaining a polished surface with this quality is about 10 times less than the time required to polish a surface of brass of the same size.

## Simple composite ferrule

Mechanical deformation is a change of geometry of the optical fiber away from a straight cylinder. Large-scale bending, or macrobending is where the radius of the curvature of the bend is very large in comparison to the core diameter. On the other hand microbends are deformations of the cylindrical core shape, which are small, compared to the fiber diameter (Ransey 1988). It is well known that mechanical deformation causes FRD by the formations of microbends in the fiber (Clayton 1989). When mounting the fiber, the appropriate epoxy and tubing should be selected, and general care must be taken to minimize mechanical stress and avoid additional FRD. Currently steel ferrules tubes are used to prepare the extremities of the optical fibers for general purposes, in test lab or even as a part of some instrument. Although it is clear that inefficiencies can result in the use of metal ferrules submitted to low temperatures. Ferrules and inserts made with the composite described here, promises to be best option to handle the ends isolated of optical fibers.

## Composite

While there is no direct evidence for the deterioration in Focal Ratio Degradation (FRD) of optical fibers in severe temperature gradients, the fiber ends inserted into metallic containment devices such as steel ferrules can be a source of stress, and hence increased FRD at low temperatures. In such conditions, instruments using optical fibers may suffer some increase in FRD and consequent loss of system throughput when they are working in environments with significant thermal gradients, a common characteristic of ground-based observatories. It is possible to use careful methodologies that give absolute measurements of FRD to quantify the advantages of using epoxy-based composites rather than metals as support structures for the fiber ends. This is shown to be especially important in minimizing thermally induced stresses in the fiber terminations. Furthermore, by impregnating the composites with small cerium oxide particles the composite materials supply their own fine polishing grit, Fig. 9, which aids significantly to the optical quality of the finished product. Different types of inserts are possible, Fig. 10, depending only on the precision of the machining.

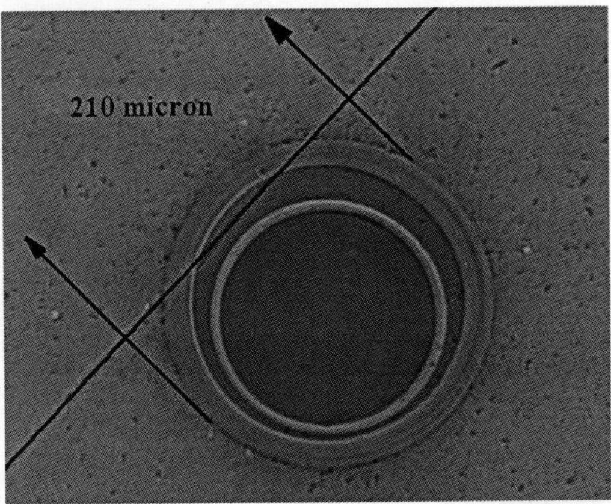


**Figure 9.**Microscopic photo of optical fiber inserted in a composite ferrule, after polishing procedure.

**Figure 10.**Inserts with optical fibers to be used in a fibers collector plate. Each insert can have several fibers.

## Microholes arrays using plates of composite

A system like that shown in section 1, Fig. 2, presented a problem in the past: This problem was the terrific facility to detach the glass substrate of the metal brass polished. Variations of temperature at long of time cause different expansion in the metal brass and the glass. After some time, the UV epoxy normally used to glue the microlens arrays with the microholes array, cannot support more the bonding between the metal brass plates due to the successive expansions and contractions caused by temperature variations. Experiments using plates made with epoxy solidified, Fig. 11 can resolve this kind of problem in the range of temperatures between –10 °C and 22 °C, typical of high altitude, ground-based observatories. Although the coefficient of thermal expansion of epoxy, around 60 x 10-6 in/in/°C, is much greater than that of brass metal, steel metal or glass, its elasticity accommodates dimensional changes thermally induced. This means less pressure on the optical fiber and consequently avoids increases in FRD associated with stress but, in principle, could be deleterious in compromising the critical positioning of fibers as the temperature varies. The ideal condition requires a material with elasticity controlled so as not to cause stress or shift the positioning of optical fiber under temperature gradients.

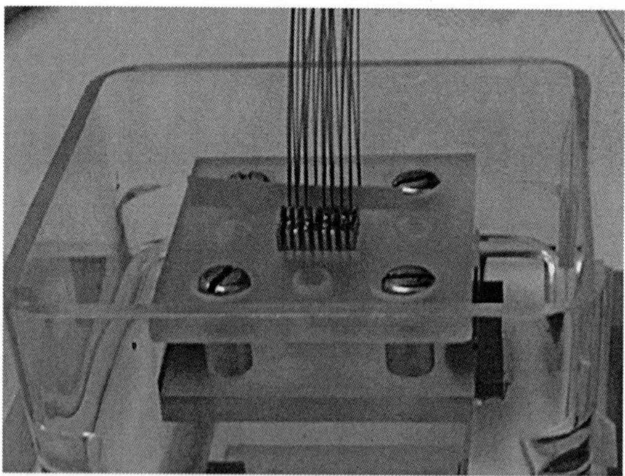

**Figure 11.**Microholes array device being prepared to be the input array of an IFU system.

Notwithstanding the characteristics of this particular composite are still under study, this material was used successfully in the construction of devices for several fibers instruments. For example, we have used this composite to construct SIFS/IFU for the SOAR telescope in Chile, (de Oliveira et al. 2010) and FRODOspec/IFU for the Liverpool Telescope, (Macanhan et al. 2006).

## Construction of microholes arrays systems

To replace the brass metal or epoxy solidified and resolve the problems presented by both materials, we have used our composite to manufacture the parts of the microholes array device. The material composite obtained is less stressed and harder than EPOTEK 301-2 being a good choice to be used in optical fibers arrays. This material certainly has a combination of the characteristics from the epoxy and the refractory material oxides. The most important consequence of this combination is a coefficient of thermal expansion hither than metal brass but shorter than a simple epoxy solidified. The exactly number will depend of the relative concentration between the refractory material oxides and epoxy.

**Figure 12.**Schematic of the composite plates set to build the entrance device of lenslet IFU.

In general, the schematic shown in Fig. 12 is the base of the entrance device of the lenslet IFU system. This device is much easier to be manufactured than the device described in section 1, Fig. 2. In fact, this new version does not require any precision in the holes confection on the composite plates. To obtain precision with the fibers position we have used a third plate called mask of precision, Fig. 13. This is a metal mask very thin obtained by a technique called electro formation. The mask obtained by this way may be configured to have holes with specifics diameters and pits, with error around 1 micron in the diameter and in the position of the holes. This technique may produce a metal nickel plate with 200 microns of thickness and the procedure is very cheap. Taking in account these facilities; the mask will define the precision of the fibers array. It is possible to obtain micro holes with the diameter exactly one or two microns larger than the diameter of the fiber used. For another hand, the diameter of the holes in the composite plates does not need to have any precision and may be much larger than the diameter of the fiber. Since that, the step holes with different diameters in the composite plates it is not more necessary, also will be not necessary to use ferrules and any kind of protection to the fibers. Eventually a device like as shown in the Fig. 13 need to be made under a microscope because the diameter of the fiber may be much small and the number of fibers involved at the assemble may be high.

**Figure 13.**Entrance device during the assembling step, where the matrixes of holes in the composite plate set and in the precision mask are populated with the optical fibers terminations.

After assembled, the precision mask is glued against the composite plate and all set is immersed in EPOTEK 301-2 following the old procedure. To obtain the maximum throughput the surface of the fibers should be polished such that they are optically flat. This is the condition to attach the microlens array against the composite plate, Fig. 14 and Fig.15.

**Figure 14.**SIFS/IFU microlens glued.

**Figure 15.**FRODOS IFU microlens glued.

The pre-polishing process starts with the removal of excess glue with 2000 grit emery paper. Initial lapping with 6 μm diamond slurry on a copper plate and a second lapping with 1 μm diamond slurry on a tin-lead plate is used until the complete removal of the precision mask. Without the metal mask, the material of the composite plate is self-abrasive enough to produce a polishing of high performance of the optical fibers on a chemical cloth. This procedure is a basic condition to attach the microlens array against the composite plate of fiber terminations.

## CHARACTERIZATION

The complete characterization of this composite may require several kinds of possible tests. However, applications with optical fibers in metrology involve analyses of displacement when the device is submitted at thermal gradients. More specifically, optical fibers arrays used in astronomic instruments need to resist low temperatures without displacement of the fiber position and without delamination problem between parts. Simples experiments show that the linear CTE assumes values between 20 and 40 x 10-6 / °C to 0 °C depending the concentration of the components. For example, a sample made with EPO-TEK 301-2, Barium oxide, Zircon oxide and Cerium oxide with proportions respectively 5:1:1:1, exhibits an α CTE around 30 x $10^{-6}$ 1/°C.

Analysis of the Absolut Transmission in samples shows clearly that optical fibers inserted in brass or steel ferrules suffer increases in FRD when submitted to low temperatures. On the hand, the FRD increase in optical fibers inserted in ferrule made from EPO-TEK or in composite materials is minimised when submitted at the same negative variation of temperature. In fact, the result predicts a loss of around 10 per cent for the brass ferrules and around 3 per cent for the steel. Although it is clear that inefficiencies can result in the use of metal ferrules submitted to low temperatures, the losses are not easily quantifiable. The reason for this is that the ratio of the outer diameter of the fiber and the inside diameter of the ferrule defines the amount of epoxy between the ferrule and fiber. In the final analysis, this represents more or less compression in the fiber when the metal is compressed during the reduction of temperature.

## Tests on fibers in an array

It is possible to do an experiment to observe the displacement of the fibers in an array of fibers constructed in the plates described in the section 4.2. An experimental array of 10 x 8 optical fibers is chosen as a representative test since it matches the base of the input array of IFUs as described before. A displacement is likely to occur with variations in temperature since the support material may suffer from some type of mechanical distortion. The experimental assembly consists of a support to hold the input array and to control thermal dissipation. A relay lens is used to project the image of the fiber array onto a CCD as a shown in the Fig. 16. The support to hold the array under examination (Fig. 17 and 18) is made of brass and had a canal for the introduction of liquid nitrogen. A continuous flux of dry nitrogen needs to be directed towards the surface of the input array to avoid condensation. Four small temperature sensors are cemented inside holes in the tested plate, close to the optical fibers. These sensors are necessary in order to test if the temperature along the plate reach thermalized state. A digital thermometer can be used to collect information from the sensors.

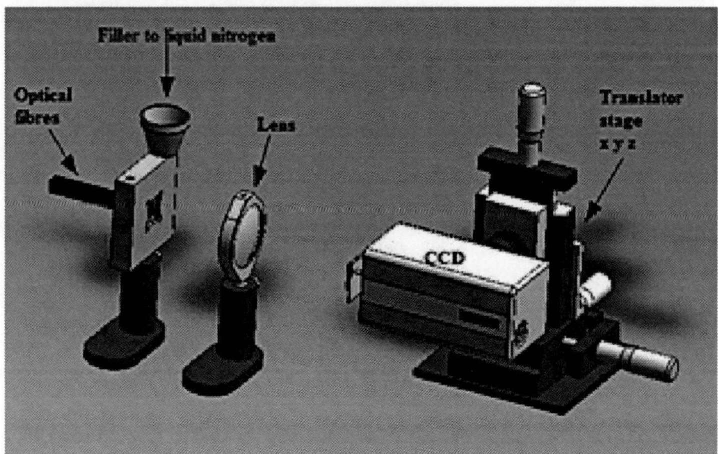


**Figure 16.**Diagram of the experimental set up to take images from the optical fiber array. The CCD is installed in a translation stage to put the image of the optical fiber illuminated exactly in the centre of the CCD plate. The holder support is used to keep the fiber plate array fixed and to control thermal dissipation.

In our experiment, another sensor was installed inside the brass support together with a special electrical resistor to allow for temperature control. The temperature of the input array was controlled over a range between 23 °C and -10 °C. Images of the illuminated optical fibers can be obtained for a set of temperatures within the allocated range. With these images it is possible to obtain information regarding the change in the position of each fiber in the array. A simple algorithm may be used to process these results.

**Figure 17.**Schematic diagram of the holder where is fixed the fiber plate array.

## Analyses of displacements of fibers in the array

An image analysis by software then determines the centroid of each bright spotlight projected by each fiber from the array on the CCD. Several images like the sample shown in the Fig. 19 are used to determine an average value in the position of each bright spot light. This associates position vectors connecting each bright spotlight with the origin. The first fiber at the top/left is used as a position reference. The variations of vector's modulus during the temperature gradient are computed to produce a graph with sub pixel precision (Neal et al. 1997). In this section we present results of tests using fiber arrays submitted to negative temperature gradients. The purpose of these tests is to analyse how much the fibers in the array may be displaced from their original position as a function of the expansion of the material during the change of temperature. Figs. 20, 21 and 22 show the behaviour of arrays with 80 optical fibers when submitted to four temperature gradients. The accuracy calculated for this experiment was less than 0.2μm and the continue curve represents a fitted function.

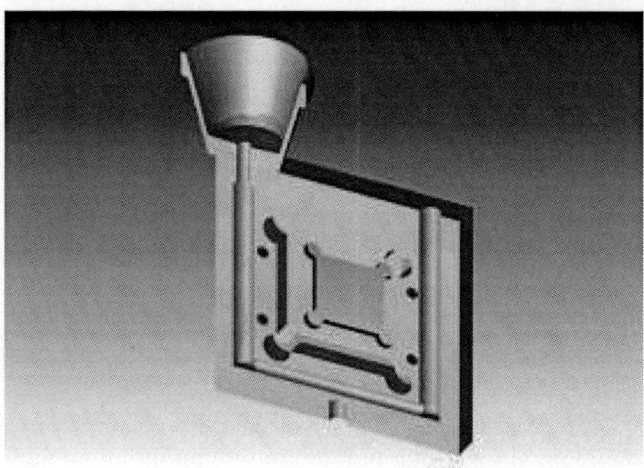

**Figure 18.**Schematic diagram of the holder showing the canal to flow the liquid nitrogen during the procedure to decrease the temperature. The holder is made of brass.

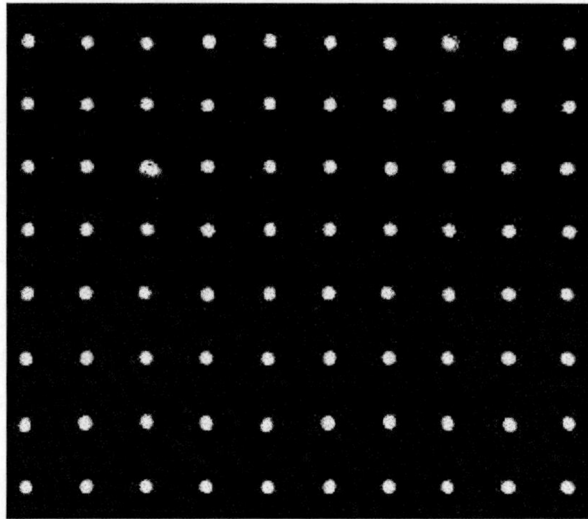

**Figure 19.**Image of the optical fibers matrix illuminated and projected on the CCD.

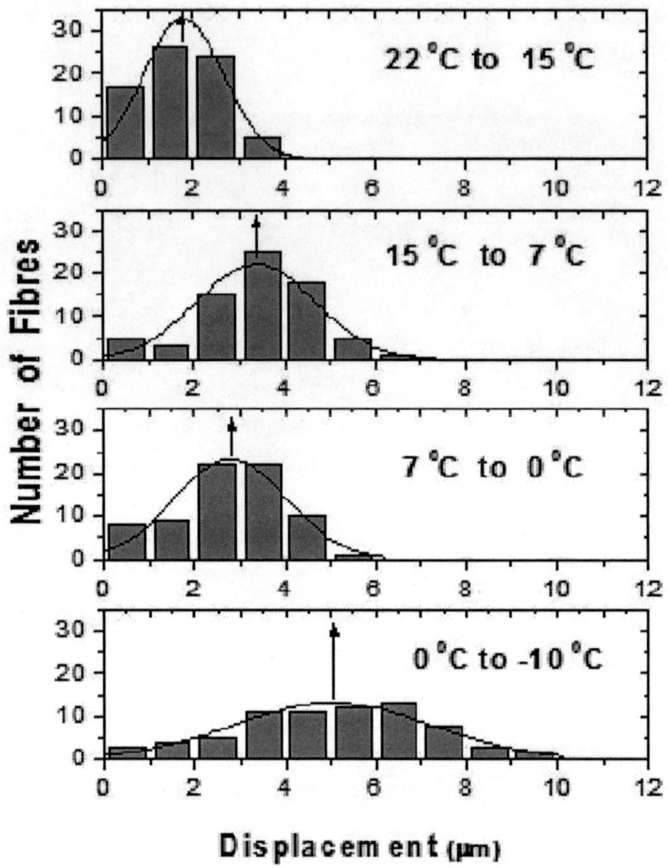


**Figure 20.**Distribution pattern of the optical fibers array constructed in metal brass plate submitted to four gradients of temperatures, 10 min each.

The bar graphs, demonstrates the distribution pattern of fiber positions as the temperature declines. It is possible to observe, the expansion of the EPO-TEK 301-2 epoxy material. The change in positions of the fibers amounts to ~12μm as the temperature approaches -10 °C. The array made of brass almost reaches this value despite having a totally different molecular structure to the epoxy. An interesting result was obtained with the optical fiber array in composite as may be observed in the Fig. 23.

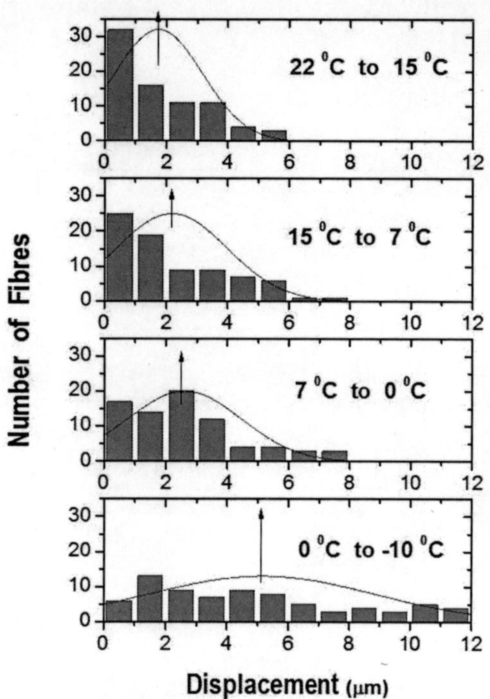


**Figure 21.**Distribution pattern of the optical fibers array constructed in epoxy EPO-TEK 301-2 submitted to four gradients of temperatures, 10 min each.

The behaviour of the composite array at low temperatures, represented in the bar graphs, is less noticeable than that obtained for the brass and epoxy arrays. In fact, when submitted to -10 °C the change of the positions at the fibers is less than 6μm. In the experimentation made with composite and brass plate samples, the temperature registered by all sensors on the plate was the same after some minutes and the error expected between the sensors would be around 0.2°C. However, we noted variations of ~1.2°C between the sensors in the experimentation with EPO-TEK plate. This may be explained by the fact that there are regions with different degrees of stress in the solidified EPO-TEK plate. These differences imply in a possible variation of thermal conductivity along the plate. As was shown in the section 2.3, stresses within solidified EPO-TEK 301-2 can be investigated with the use of the photo elasticity method whereby stress-induced birefringence is measured using polarised light. In fact it is quite common to observe static stress in plates of EPO-TEK solidified.

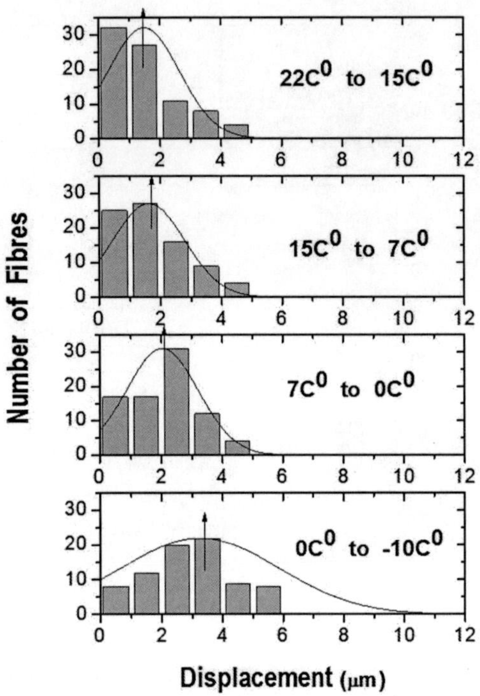


**Figure 22.**Distribution pattern of the fibers array in composite submitted to 4 gradients of temperatures, 10 min each.

The final conclusion for this experiment is that the composite epoxy material shows significant improvement and, in fact, has an even better performance than the brass or epoxy solidified. The chosen composite material (EPO-TEK 301-2 + zirconium oxide) retains the beneficial bonding properties of the epoxy while avoiding its thermal displacement properties.

## FRD in optical fibers samples

The mode dependent loss mechanisms are the causes of focal ratio degradation (FRD) in optical fibers, and are not often addressed by manufacturers. Mode dependent losses can be divided into two basic mechanisms. The first is waveguide scattering, which causes transfer of energy into loss modes by variations of the core diameter along the length of the fiber. The second is mechanical deformation. Mechanical deformation is a change of the geometry of the fiber away from a straight cylinder. Large scale bending, or macrobendings, is where the radius of curvature of the bend is very large in comparison to the core diameter. On the other hand, microbends are deformations of the cylindrical core shape, which are small, compared to the fiber diameter (Ransey 1988). It is well known that mechanical deformation causes FRD by the formation of microbends in the fiber (Clayton 1989). FRD is a non-conservation of *étendue* (or optical entropy) such that the focal ratio is broadened by propagation in the fiber. When mounting the fiber, the appropriate epoxy and, tubing should be selected and general care must be taken to minimise mechanical stress and avoid additional FRD.

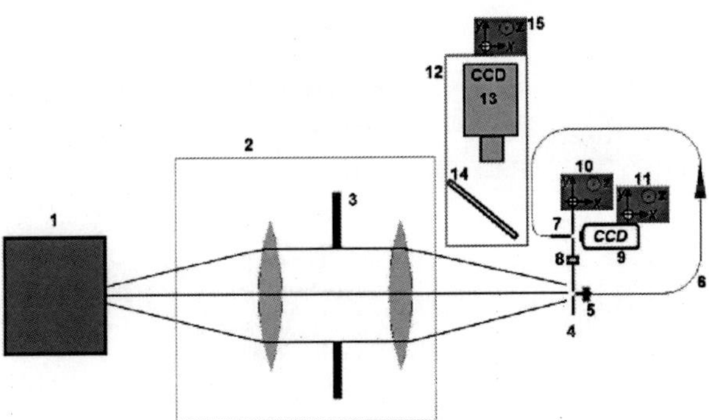


**Figure 23.**Diagram of the apparatus used to measure FRD – 1, light source, band pass filter and light diffuser; 2, telecentric optical system with unit magnification; 3, adjustable iris diaphragm; 4, alignment plate with a pinhole and both extremities of the tested fiber; 5, *peltie*r device connected with the entrance of the optical fiber; 6, optical fiber; 7, exit of the optical fiber; 8, pinhole; 9, CCD; 10, xyz translation stage; 11, xyz translation stage; 12, microscope system; 13, CCD/lens; 14, beam splitter; 15, xyz translation stage.

To measure the FRD properties of an optical fiber it is necessary to illuminate the test fiber with an input beam of known focal ratio. Then the output beam can be measured and compared with the input beam from a pinhole with the same diameter as the fiber core to determine the amount of FRD produced by the test fiber. The result is a plot of absolute transmission against output focal ratio. The experimental apparatus used to achieve this is illustrated in Fig. 23.

Illumination is provided by a 1-to-1 telecentric optical system that produces an image from an extensive uniformly illuminated source. This source is fed by a stabilized halogen lamp and has a band pass filter to provide light at 525nm, and filter's bandwidth of 100nm. An iris diaphragm placed in the collimated beam can be used to select the input focal ratio. A microscope with a CCD and beam splitter, monitored by a TV may be inserted between the pinhole/ fiber plane, to be sure that the pinhole or the test fiber occupies the same position. To ensure accurate alignment of the fiber with the optical axis of the camera, the fiber is mounted in a tip-tilt translation stage. To begin the experiment, the pinhole device and the CCD are positioned to give us a reference image. In the test sequence, the pinhole is replaced with the entrance of the test fiber and the CCD is illuminated by the exit of the test fiber to give a projected image of the fiber. A distance of 9 mm between the CCD and the pinhole (or the entrance of the fiber in test) was determined as the best position to obtain images for optimal analysis. Background exposures are necessary for subtraction from the test exposures to remove the effects of hot pixels and stray light. In our experiments all fibers were tested at wavelength of 525nm, (defined using a Schott glass VG14 colour filter, ± 50 nm filter's bandwidth).

## Reduction software

We have developed a custom software package (DEGFOC 3.0) to reduce the fiber images and to obtain throughput energy curves. This software works with PC microcomputers in a WINDOWS environment. We found this to be an effective solution for use in the optical laboratory environment allowing for ease of analysis. The DEGFOC 3.0 package gives curves of enclosed energy as is shown in the Fig. 24 with the option to save the result in ASCII format to be used in any graphic software, (eg: ORIGIN).

Fiber throughputs are automatically determined as a function of output focal ratio. The first step is an estimation of the background level to be subtracted from the test exposures to remove the effects of hot pixels and stray light. The software then finds the image centre by calculating the weighted average of all pixels. It associates a radius with each pixel and calculates the eccentricity that, in the ideal case, should be zero. Our target here is to obtain the absolute transmission of the fiber at a particular input f-ratio. After establishing the distance between the fiber test and the CCD, the software defines concentric annuli centred on the fiber image. These are then used to define the efficiency over a range of f-numbers at the exit of the fiber, where each f-number value contains the summation of all energy emergent from the fiber. Each energy value is calculated by the number of counts within each annulus divided by total number of counts from the pinhole images. The limiting focal ratio that can propagate in the tested fiber is approximately f/2.2. Therefore we have defined f/2 to be the outer limit of the external annulus within which all of the light from the test fiber will be collected. The corresponding diameters of the annulus are converted to output focal ratios, multiplying them by the appropriate constant given by the distance between the fiber output end and the detector.

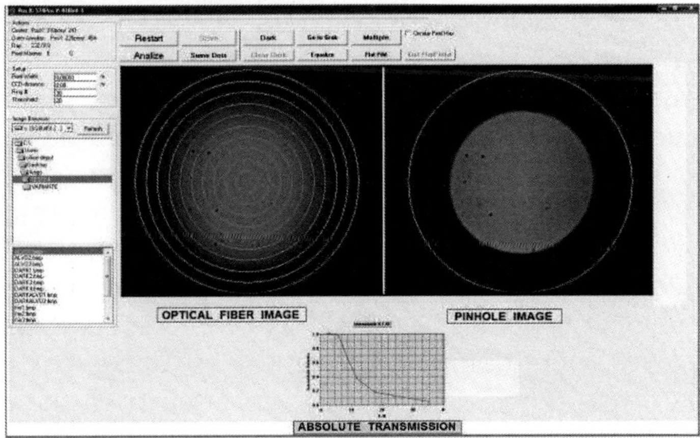


**Figure 24.**Print screen of the windows to the DEGFOC software.

## Temperature gradient & FRD in optical fibers

In this experiment we have controlled the temperature of

samples between –10 °C and 22 °C, typical of high altitude, ground-based observatories. To achieve this variation we have used a *Peltier* device coupled with a temperature sensor connected to an electronic controller. The end of the test fiber is placed in contact with the *Peltier* plate by a support, and to avoid problems with water condensation at low temperatures, the test ferrule is installed inside a plastic container with a glass window. A positive pressure of nitrogen gas is maintained using a flexible tube from a gas source. With these experimental arrangements it is possible to obtain images of the optical fibers with one of extremities inside a ferrule experiencing low temperatures without water condensation. This avoids the formation of ice at the end of the fiber that could attenuate the light at its termination and contaminate the results. Our aim is to measure the effect of constriction of the ferrule on the optical fiber caused by the gradient in temperature.

Plots of absolute transmission versus output focal ratio for three samples in four configurations are presented here. We have plotted graphs with the extremes curves obtained at room temperature of 23 °C and at -10 °C after a time interval of 30 min. chosen to stabilize the thermal effects between one measurement and next. The throughput graph obtained from one fiber with brass ferrule, in dry atmosphere, is shown in Fig. 25. These results show an increase in FRD when the brass ferrule experiences a cold temperature. The total variation observed in the hatched area is very strong and diminishes as the output focal ratio of the fiber is increased. An analysis of the results demonstrates that the loss of light at F/2.3 would be around 10 per cent. This degradation is caused, presumably, by the contraction of the brass ferrule with decreasing temperature causing compressive stress of the ferrule on the fiber. The error bars, of ± 1 per cent, together the average curves, were defined after repeating each experimentation at least six times. Some experimental uncertainty in the control of temperature causing small variations in the compression force on the ferrule/fiber and consequently cause small variations in the throughput of the sample. However is evident the presence of some dissipative process, which changes the borderline of the stress during the variations of temperature. Taking in account that the experiment is made with input focal ratio around the Numerical Aperture of the fiber (F/2.27) there is the possibility that it is changing because the stress.

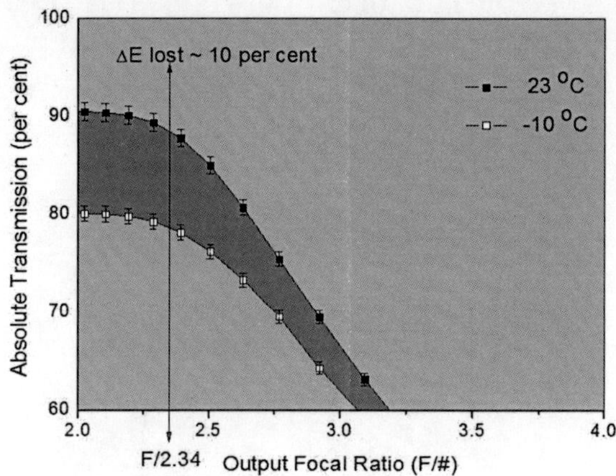


**Figure 25.**Performance of the optical fiber using brass ferrule. The ferrule was submitted to a negative temperature gradient of 23 °C in a dry atmosphere. The gradient was obtained, reducing the temperature, 23 °C to -10 °C, in 30 min of interval time. Two extremes curves were measured in this interval, producing the hatched area.

Such effects are critical to the design and implementation of fiber spectrographs. These results imply serious restrictions in the use of metal ferrules for optical fibers operating in ambient conditions that experience large changes of temperature typical of many observatories both during the night and throughout the year.

The throughput for samples with fibers inserted into epoxy ferrules and composite ferrules, in dry atmosphere, are shown for comparison in Figs. 26 and 27 using the same experimental procedures. Both graphs, present a very similar curves, with loss of light at F/2.34 around 2 per cent to the epoxy ferrule and 1 per cent to the composite ferrule. It seems that the loss of energy through stress-induced FRD effects is significantly less than that observed with the metal ferrule samples. The similarity of the epoxy and composite results imply that we are seeing similar effects due to the similar structure of both materials. In fact the composite material uses the same epoxy as a substrate. A natural compression happens during the cooling process, but does not produce a compressive stress of the brass ferrule on the fiber. The elastic properties of the epoxy may neutralize the mechanical stress on the fiber during the contraction process. The same error bars, of ± 1 per cent, were obtained after six repetitions of the experiment.

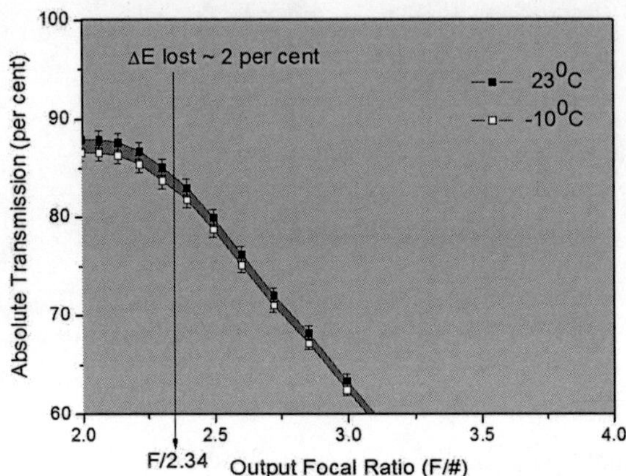


**Figure 26.**Performance of the optical fiber using epoxy ferrule. The ferrule was submitted to the negative temperature gradient following the same conditions of the experimentation using metal ferrule.

In general, the variation in the FRD results obtained with different samples from the same optical fibers is ±~1 per cent because the noise of the measurements. Analyse of the throughput curves obtained at room temperature from the epoxy and composite ferrules is ~ 3 per cent less on average when comparing the same curve obtained from the metal ferrules samples. The explanation for this difference may be in the aging process of the epoxy and composite ferrules. Both samples were submitted to six thermal cycles, between 50 °C and -20 °C after machining to avoid anomalous results during the experimentations. However, this procedure may increase the intrinsic FRD of the fiber given that the material structure of the ferrule may suffer accommodation pressing the fiber extremity. On the other hand, small differences of size in the hatched area of lost energy between similar samples could be expected. Differences like that would be explained by the difficulty to quantify the total length of the fiber immersed in epoxy inside the ferrule. The procedure of inserting fiber and epoxy into the ferrule is virtually handmade. There is no way to accurately control the amount of epoxy into opaque ferrules, because it is not possible to visualize the level of epoxy. Exception perhaps for polished quartz ferrules. Obviously, the length of fiber immersed in epoxy defines the length that would be submitted at the stress from the ferrule contraction.

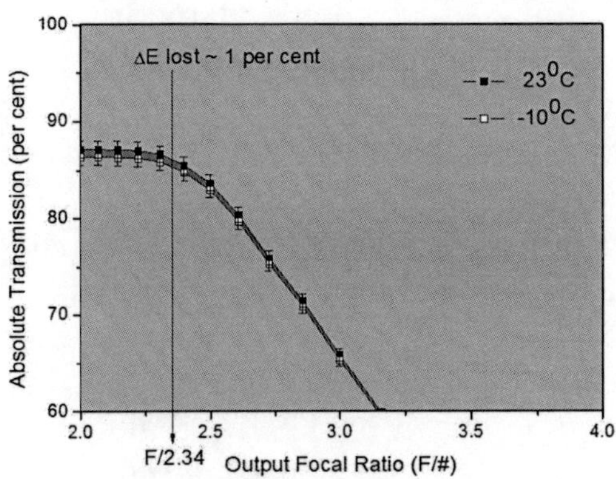


**Figure 27.**Performance of the optical fiber using composite ferrule. The ferrule was submitted to the same negative temperature gradient of the anterior experimentation using metal ferrule, steel ferrule and epoxy ferrule.

## Polishing substrate

There are two ways of surface preparations in optical fibers, cleaving and polishing. In general applications directed to scientific instrumentation require optical fibers with extremities polished. This is the way where it is possible to optimize the spot light from the optical fiber. Furthermore, all fiber connectors require polishing and high performance may be reached with special machines and dedicated procedures. Currently, polishing procedures to optical fibers are based on very delicate glass paper or lapping discs soaked in abrasive liquid solutions. Often, this kind of liquid abrasive is very expensive taking in account your composition based in sophisticated chemistry keeping micro diamonds in suspension.

Other options, considers abrasive silica and aluminium oxide mixed with oil solution or water solution. A very interesting application for the composite described here is its use as a high-performance abrasive disc to polish optical fibers.

## Abrasive discs of composite to polish optical fibers

It is possible to fabricate composite discs, controlling the abrasive capacity through the correct choice of oxide and quantity mixed with epoxy. There are several manufacturers of oxide refractory with high purity such that it is possible to compose a complete grid of polishing discs. We can consider two major advantages in the use of polishing discs manufactured with composite: The first lies in the fact that the entire polishing process can be done using distilled water only. The second advantage is that after the polishing procedure, the disc can be restored to its original flatness and completely cleaned by machining process. The efficiency of this composite disc to polish optical fibers is based in the fact that some of the oxides of the mixture are naturally abrasives. Materials like cerium oxide or silica oxide can be prepared in liquid solutions abrasives and has been used for a long time in polishing procedures of lenses and other optical devices. Discs of composite can be made in any size and can easily be adapted in the rotation device of polishing machine. Fig. 28 shows an array of optical fibers polished using discs of composite.

**Figure 28.**Microscopic photo of part of the optical fibers array, after polishing procedure, using discs of composite containing cerium oxide.

## Conclusion

The motivation of this work was to test the performance of optical fibers inserted in ferrules made with different materials at low temperatures. The problem of finding a material best suited to securing fibers for astronomical spectrographs to cope with thermal stresses, FRD minimization and the need to achieve adequate polishing finish led us to investigate the use of composite materials. As has already been demonstrated, epoxies can be used not only as a means of holding fibers within structures (slit blocks, fiber arrays etc.) but also as a material to fabricate the structures themselves. The properties that require investigation in this context are CTE matching, machinability, bonding to glass and ease of polishing. In this context we have made several samples to evaluate FRD performance and position displacement of the inserted fibers when submitted to low temperatures.

## ACKNOWLEDGEMENT

This work was financially supported by the FAPESP project no. 1999/03744-1 and CNPq project 62.0053/01-1- PADCT III/ Milenio. We wish to thank the staff of the Laboratório Nacional de Astrofísica/ MCT.

## REFERENCES

1. C. A. Clayton, 1989The Implications of Image Scrambling and Focal Ratio Degradation in Fiber Optics on the Design of Astronomical Instrumentation, Astronomy and Astrophysics, 2131-2April 1989), 5025150004-6361
2. A. C. de Oliveira, et al.2002The Eucalyptus Spectrograph, Proceedings of SPIE Instrument Design and Performance for Optical/Infrared Ground-based Telescopes, 14171428Waikoloa, Hawaii, USA, August 25-28, 2002
3. A. C. de Oliveira, et al.2005Studying Focal Ratio Degradation of Optical Fibers with a Core Size of 50µm for Astronomy, Mon. Not. R. Astron. Soc., 3563October 2004), 107910870035-8711
4. A. C. de Oliveira, et al.2010The SOAR Integral Field Unit Spectrograph Optical Design and IFU Implementation, Proceedings of SPIE

Modern Technologies in Space- and Ground-based Telescopes and Instrumentation, 77394S773941S-12, San Diego, California, USA, June 27, 2010

5. L. W. Ransey, 1988Focal Ratio Degradation in Optical Fibers of Astronomical Interest, In: Fiber Optics in Astronomy, Samuel C. Barden, 2640Astronomical Society of the Pacific, 0-93770-720-1Francisco, California, USA
6. V. B. P. Macanhan, et al.2006FRODOSPEC Integral Fiber Unit, Proceedings of SAB XXXII Reunião da Sociedade Astronômica Brasileira, 194195Atibaia, São Paulo, Brasil, August 3, 2006

# Citations

## CHAPTER 1

Paulo Cesar Plaisant Junior, Flávio Luiz de Silva Bussamra, Francisco Kioshi Arakaki, Finite element procedure for stress amplification factor recovering in a representative volume of composite materials doi: 10.5028/jatm.2011. 03033911

## CHAPTER 2

Roelof Marissen, Design with Ultra Strong Polyethylene Fibers, doi:10.4236/msa.2011.25042

## CHAPTER 3

Effects of rc beams reinfocement using near surface mounted reinforced frp composites slobodan ranković1, radomir folić2, marina mijalković1, doi: 10.2298/fuace1002177r

## CHAPTER 4

QUINI, Josue Garcia and MARINUCCI, Gerson.Polyurethane structural adhesives applied in automotive composite joints. Mat. Res. [online]. 2012, vol.15, n.3, pp. 434-439. Epub May 08, 2012. ISSN 1516-1439. http://dx.doi.org/10.1590/S1516-14392012005000042.

## CHAPTER 5

M. Bocciolone, M. Carnevale, A. Collina, N. Lecis, A. Lo Conte, B. Previtali , C.A. Biffi, P.Bassani, A. Tuissi , Application of martensitic SMA alloys as passive dampers of GFRP laminated composites, DOI: 10.3221/IGF-ESIS.23.04.

## CHAPTER 6

Gabriel oprişan*, vlad munteanu, nicolae ţăranu and alina lazăr fiber reinfoced polymer used for flooding protection of engineering structures made of rc and brick masonry, issn: 1224-3884.

## CHAPTER 7

D. Murray and O. Myers, "Modeling Fiber Composites during the Cure Process for Piezoelectric Actuation,"World Journal of Mechanics, Vol. 3 No. 1, 2013, pp. 26-42. doi: 10.4236/wjm.2013.31002.

## CHAPTER 8

Gaurav Agarwal Amar Patnaik and Rajesh Kumar Sharma Thermo-mechanical properties of silicon carbide filled chopped glass fiber reinforced epoxy composites doi: 10.1186/2008-6695-5-21

## CHAPTER 9

H.P.S. Abdul Khalil, M. Jawaid, A. Hassan, M.T. Paridah and A. Zaidon (2012). Oil Palm Biomass Fibres and Recent Advancement in Oil Palm Biomass Fibres Based Hybrid Biocomposites, Composites and Their Applications, Prof. Ning Hu (Ed.), ISBN: 978-953-51-0706-4, InTech, DOI: 10.5772/48235. Available from: http://www.intechopen.com/books/composites-and-their-applications/oil-palm-biomass-fibres-and-recent-advancement-in-oil-palm-biomass-fibres-based-hybrid-biocomposites

## CHAPTER 10

Alexander Horoschenkoff and Christian Christner (2012). Carbon Fiber Sensor: Theory and Application, Composites and Their Applications, Prof. Ning Hu (Ed.), ISBN: 978-953-51-0706-4, InTech, DOI: 10.5772/50504.

## CHAPTER 11

Antonio C. de Oliveira and Ligia S. de Oliveira (2012). Composite Material and Optical Fibers, Composites and Their Applications, Prof. Ning Hu (Ed.), ISBN: 978-953-51-0706-4, InTech, DOI: 10.5772/47876.

# INDEX

## F



## G



## H



## I



## K



## L



## M



## N



## O



## P



## Q

## R



## S



## T



## U



## V



## W